ROUTLEDGE L
 I

Volume 14

TOPICS IN MODERN LOGIC

TOPICS IN MODERN LOGIC

D. C. MAKINSON

Routledge
Taylor & Francis Group

LONDON AND NEW YORK

First published in 1973 by Methuen & Co Ltd

This edition first published in 2020
by Routledge
2 Park Square, Milton Park, Abingdon, Oxon OX14 4RN

and by Routledge
52 Vanderbilt Avenue, New York, NY 10017

Routledge is an imprint of the Taylor & Francis Group, an informa business

British Library Cataloguing in Publication Data
A catalogue record for this book is available from the British Library

ISBN: 978-0-367-41707-9 (Set)
ISBN: 978-0-367-81582-0 (Set) (ebk)
ISBN: 978-0-367-42051-2 (Volume 14) (hbk)
ISBN: 978-0-367-42622-4 (Volume 14) (pbk)
ISBN: 978-0-367-85415-7 (Volume 14) (ebk)

Publisher's Note
The publisher has gone to great lengths to ensure the quality of this reprint but points out that some imperfections in the original copies may be apparent.

Disclaimer
The publisher has made every effort to trace copyright holders and would welcome correspondence from those they have been unable to trace.

Topics in Modern Logic

D. C. MAKINSON

Methuen & Co Ltd
11 New Fetter Lane
London EC4P 4EE

First published 1973
by Methuen & Co Ltd,
11 New Fetter Lane, London EC4P 4EE
© D. C. Makinson
Typeset in Great Britain by
William Clowes & Sons Ltd, London, Colchester and Beccles
Printed in Great Britain by Butler & Tanner Ltd, Frome

SBN 416 77440 7 Hardback
SBN 416 78100 4 Paperback

Distributed in the U.S.A. by
HARPER & ROW PUBLISHERS, INC.
BARNES & NOBLE IMPORT DIVISION

Contents

Preface

This book is directed to the undergraduate student of philosophy who wishes to study modern logic beyond the most elementary level, but whose background in mathematics is very limited. It is not intended as a first introduction to logic. The student should already have behind him something like a term's work in the subject, and should have a rough grasp of ideas such as those of a truth-functional operator, tautology, predicate, relation, and quantifier.

The author has tried to strike a balance between material of a philosophical and a formal kind, and to do this in a way that will bring out the connections between the two.

On the formal side, particular care is given to providing the basic tools from arithmetic and set theory that are needed for the study of systems of logic, setting out a maximality argument for obtaining completeness results in two, three, and four valued logic, explaining the concepts of freedom, bondage, and regular substitution in quantificational logic, describing the intuitionistic interpretation of each of the logical operators, and setting out Zermelo's axiom system for set theory.

On the philosophical side, particular attention is given to the problem of entailment, the import of the Löwenheim-Skolem theorem, the limits to the expressive capacities of quantificational logic, the ideas underlying intuitionism, the relationship between logic and set theory, and the nature of set theory itself. There is also a digression into heuristic, with a discussion of the difference between working forwards and working backwards, or as Descartes would call them, the synthetic and analytic modes of inquiry.

The book is brief, and there is much in logic that it does not cover. It is intended to provide the basic material for a term's work.

There are exercises within the text, set out alongside the theoretical ideas that they involve, and selected so as to be well within the reach of a student who lacks a mathematical background. Solutions have also been provided for some of the exercises. Finally, there is a guide to further reading, for the deeper study of questions arising in the text.

The material was originally used in courses at the American University of Beirut, Lebanon. Thanks are due to Dr J. E. J. Altham and Professor H. W. Johnstone, Jr. for their remarks on an earlier version. Miss Suzie Khatchadourian and Miss Maleeha Suidan helped with the typing.

D.C.M.

American University of Beirut
February 1972

1 Some aspects of truth-functional logic

Outline

We begin by setting out an axiom system for the relation of tautological implication. We then give some examples of the derivation of theses in the axiom system, and digress to discuss the difference between working forwards and working backwards, both in the search for derivations and in problem solving in general. We describe some of the principal tools, drawn from arithmetic and the theory of sets, that are needed in the study of systems of logic. Finally, we use these tools to show that the axiom system is both sound and complete with respect to the relation of tautological implication.

1 An axiomatization of the relation of tautological implication

We shall assume that the reader has some acquaintance with the formulae of propositional logic. They are constructed from propositional letters $p, q, r, \ldots$ by means of operators such as negation, conjunction, disjunction, material implication, and material equivalence, which we shall write respectively as $\neg, \wedge,$ $\vee, \supset,$ and $\equiv$. We recall that of these operators, some are expressible in terms of others. For example we may take the operators $\neg, \wedge,$ and $\vee$ as primitive and define the others in terms of them, considering $\alpha \supset \beta$ as an abbreviation for, say, $\neg(\alpha \wedge \neg\beta)$ and regarding $\alpha \equiv \beta$ as an abbreviation for $(\alpha \supset \beta) \wedge (\beta \supset \alpha)$. Indeed, it is even possible to take the reduction further, but we shall find it convenient for some later developments to take all three of $\neg, \wedge,$ and $\vee$ as primitive operators.

Exercise 1   Indicate two ways in which the reduction of primitive operators might be carried further.

We recall some of the fundamental definitions of truth-functional logic. A formula is said to be a *tautology* if it receives the value 'true' for every assignment of truth values to its propositional letters. Here the word 'iff', which we shall use frequently, is merely a shorthand for 'if and only if'. A formula is said to be a *countertautology*, or in the terminology of some authors a selfcontradiction, iff it receives the value 'false' for every assignment of truth values to its propositional letters. A formula is said to be *contingent* iff it is neither a tautology nor a countertautology. And finally, we say that one formula *tautologically implies* another iff there is no assignment of truth values to propositional letters upon which the first formula comes out true and the second comes out false.

Exercise 2   Let α be any formula. Verify that α is a tautology iff $\neg\alpha$ is a countertautology. Also verify that α is contingent iff $\neg\alpha$ is contingent.

Exercise 3   Let α and β be formulae. Verify that α tautologically implies β iff the formula $\alpha \supset \beta$ is a tautology.

In this chapter we shall focus attention upon the relation of tautological implication. Our first problem will be a formal one: to axiomatize the relation. In other words we would like to pick out a few simple examples of tautological implication, and a few simple rules for generating new cases of tautological implication from old ones, in a way that will satisfy two conditions:

(1) Everything derivable from the chosen examples by means of the chosen rules really is a case of tautological implication;

(2) Every case of tautological implication can be derived, in a short or a long time, from the chosen axioms by means of the chosen rules.

Clearly, these two conditions are converses of each other. It is not obvious in advance that both objectives can be attained, but it turns out that in fact they can. Indeed there are many axiomatizations that do the job, and we shall choose the following one.

Axiom schemes

$\alpha \wedge \beta \to \alpha$ $\qquad\qquad\qquad$ $\alpha \to \alpha \vee \beta$

$\alpha \wedge \beta \to \beta$ $\qquad\qquad\qquad$ $\beta \to \alpha \vee \beta$

$\neg\neg\alpha \to \alpha$ $\qquad\qquad\qquad$ $\alpha \to \neg\neg\alpha$

$\qquad\alpha \wedge (\beta \vee \gamma) \to (\alpha \wedge \beta) \vee (\alpha \wedge \gamma)$

$\qquad\qquad(\alpha \vee \beta) \wedge \neg\alpha \to \beta$

Derivation rules

From $\alpha \to \beta$ and $\beta \to \gamma$ to $\alpha \to \gamma$

From $\alpha \to \beta_1$ and $\alpha \to \beta_2$ to $\alpha \to (\beta_1 \wedge \beta_2)$

From $\alpha_1 \to \beta$ and $\alpha_2 \to \beta$ to $(\alpha_1 \vee \alpha_2) \to \beta$

From $\alpha \to \beta$ to $\neg\beta \to \neg\alpha$

Some explanation is needed of the concept of an axiom scheme. Consider for example the first scheme, $\alpha \wedge \beta \to \alpha$. In setting this down we mean to indicate that infinitely many expressions of that form are counted as axioms. For *every* choice of formulae α and β, the resulting expression covered by the scheme $\alpha \wedge \beta \to \alpha$ is counted as an axiom. Thus if we choose α to be the propositional letter p and choose β to be the formula $q \vee \neg p$, then the resulting expression $p \wedge (q \vee \neg p) \to p$ is an axiom, and is said to be an *instance* of the first axiom scheme. Again, if we choose α to be the formula $q \vee p$, and choose β to be the formula $r \wedge s$, then the resulting expression $(q \vee p) \wedge (r \wedge s) \to q \vee p$ is an axiom, and is also an instance of the first axiom scheme.

Exercise 4 List five further instances of the axiom scheme $\alpha \wedge \beta \to \alpha$. Find an expression that is simultaneously an instance of the axiom scheme $\alpha \wedge \beta \to \alpha$ and an instance of the axiom scheme $\alpha \wedge \beta \to \beta$. Find an expression that is simultaneously an instance of the axiom schemes $\alpha \wedge \beta \to \beta$ and $(\alpha \vee \beta) \wedge \neg\alpha \to \beta$. Is there any expression that is simultaneously an instance of the schemes $\alpha \wedge \beta \to \alpha$ and $\alpha \to \alpha \vee \beta$?

Each axiom contains a single arrow, flanked on left and right by formulae. This arrow should not be conflated with the hook sign. The hook is used as an operator, building formulae out of formulae, and is understood as representing material implication. The arrow is being used to represent a certain relation between formulae which, we shall show, coincides with tautological implication.

Each axiom scheme and derivation rule is a well known personality in logical life. The topmost schemes tell us that a conjunction implies each of its conjuncts, and that a disjunction is implied by each of its disjuncts. Next we have the principles of double negation. The next line gives us one of the principles of distribution, telling us that conjunction may be distributed over disjunction. Note that the converse principle

$$(\alpha \wedge \beta) \vee (\alpha \wedge \gamma) \to \alpha \wedge (\beta \vee \gamma)$$

has not been included in the list of axiom schemes, nor have the two dual principles

$$\alpha \vee (\beta \wedge \gamma) \to (\alpha \vee \beta) \wedge (\alpha \vee \gamma)$$
$$(\alpha \vee \beta) \wedge (\alpha \vee \gamma) \to \alpha \vee (\beta \wedge \gamma).$$

There is nothing objectionable about these distribution principles, and they all correspond to genuine tautological implications. But if we take only our given distribution principle from among them as an axiom scheme, then the remainder can all be derived using the derivation rules.

The final axiom scheme, $(\alpha \vee \beta) \wedge \neg \alpha \to \beta$ is known as the principle of disjunctive syllogism. It is rather inconspicuous, but as we shall later see, leads to some curious results.

The derivation rules also express familiar principles. The first lays down the transitivity of the implication relation. The second authorizes conjoining the consequents of a single antecedent, whilst dually the third authorizes disjoining the antecedents of a single consequent. The final derivation rule expresses the principle of contraposition. The basic difference between the axiom schemes and the derivation rules is that the former state, quite categorically, that certain formulae imply others, whilst the latter tell us merely that *if* such-and-such implies so-and-so, *then* such another implies another so.

We shall say that an expression is a *thesis* of the axiom system if it is either an instance of an axiom scheme or can be obtained from instances of axiom schemes by means of a finite number of applications of the derivation rules. Thus, for example, any expression of the form $\alpha \rightarrow \alpha$ is a thesis, since it can be obtained as follows:

(1) $\alpha \rightarrow \neg\neg\alpha$ axiom
(2) $\neg\neg\alpha \rightarrow \alpha$ axiom
(3) $\alpha \rightarrow \alpha$ from (1) and (2) by rule of transitivity.

One way of understanding a structure is by playing with it. We can play with an axiom system by showing that various expressions are theses of it, as in the following exercises.

Exercise 5   Let α and β be any formulae. Show that each of the following 'commutation principles' is a thesis of the axiom system. Set your derivation out in a form similar to that given above.

$$\alpha \wedge \beta \rightarrow \beta \wedge \alpha$$
$$\alpha \vee \beta \rightarrow \beta \vee \alpha$$

Exercise 6   Let α and β be formulae. Show that each of the following 'association principles' is a thesis of the axiom system:

$$\alpha \wedge (\beta \wedge \gamma) \rightarrow (\alpha \wedge \beta) \wedge \gamma \qquad (\alpha \wedge \beta) \wedge \gamma \rightarrow \alpha \wedge (\beta \wedge \gamma)$$
$$\alpha \vee (\beta \vee \gamma) \rightarrow (\alpha \vee \beta) \vee \gamma \qquad (\alpha \vee \beta) \vee \gamma \rightarrow \alpha \vee (\beta \vee \gamma).$$

Exercise 7   Let α and β be any formulae. Show that each of the following 'absorption principles' is a thesis of the axiom system:

$$\alpha \wedge (\alpha \vee \beta) \rightarrow \alpha \qquad \alpha \rightarrow \alpha \wedge (\alpha \vee \beta)$$
$$\alpha \vee (\alpha \wedge \beta) \rightarrow \alpha \qquad \alpha \rightarrow \alpha \vee (\alpha \wedge \beta).$$

At this point it may be instructive to digress from questions of pure logic, and turn for a moment to heuristics. When tackling exercises such as those above, it is often useful to work backwards rather than forwards. In general, to work forwards is to begin by focusing attention on what is already given, and to build a bridge from it towards what is desired. To work backwards, on the other hand, is to follow the opposite procedure: it is to begin by attending to what is desired, and trying to retrace a bridge back towards what is already given. Thus in the present situation, working forwards amounts to focusing attention on the axioms of the system and on the expressions that have already been shown to be theses, and building a derivation that will carry us towards the expression that we want to show to be a thesis. Working backwards amounts to examining the expression that we want to show to be a thesis, and trying to retrace, in reverse order, the likely steps of a derivation. The difference between the two procedures lies in the way in which we search for a derivation, not in the kind of derivation sought for.

An example will make the matter clearer. Suppose that we want to show that all expressions of the form $(\alpha \wedge \beta) \vee (\alpha \wedge \gamma) \to \alpha \wedge (\beta \vee \gamma)$ are theses of the axiom system. This is not the same as the axiom scheme for distribution, but is instead its converse. Now if we were to work forwards, we would begin by asking ourselves how we could set off from the axioms in the direction of the desired expression, and it is rather difficult to think of an appropriate path to follow. But if we were to work backwards, we would begin by asking ourselves what could be the *next to last* step in a derivation of the desired expression. Now as the consequent of the desired expression is a conjunction, we know that in view of the rule for conjoining consequents, it will be enough if we can show that each of the expressions

$$(\alpha \wedge \beta) \vee (\alpha \wedge \gamma) \to \alpha$$
$$(\alpha \wedge \beta) \vee (\alpha \wedge \gamma) \to \beta \vee \gamma$$

is a thesis. But each of these expressions has a disjunctive antecedent. So in view of the rule for disjoining antecedents, it will be enough to show that each of the following expressions is a thesis:

$$\alpha \wedge \beta \to \alpha \qquad \alpha \wedge \gamma \to \alpha$$
$$\alpha \wedge \beta \to \beta \vee \gamma \qquad \alpha \wedge \gamma \to \beta \vee \gamma.$$

Now clearly the expressions on the upper line are axioms of the system, and those on the lower line are very easily derived. Thus the expression originally considered is a thesis, and the problem is solved.

Exercise 8   Rearrange the above material in the form of an ordered sequence that begins with axioms and whose last item is the expression that was shown to be a thesis.

Exercise 9   Let α, β and γ be any formulae and consider the expression $\alpha \vee (\beta \wedge \gamma) \to (\alpha \vee \beta) \wedge (\alpha \vee \gamma)$. Work backwards to

discover a derivation of the expression, and then work forwards
to exhibit the derivation as an ordered sequence that begins with
axioms and ends with the desired expression.

Clearly, the idea of working backwards is of great generality, and
applies to problems beyond our axiom system. The method of
working backwards was described, in the context of problems of
geometry, by the Greek mathematician Pappus. It was revived by
Descartes, who referred to it as the 'analytic' mode of inquiry in
contrast with the method of working forwards, which he called the
'synthetic' mode of inquiry. Descartes consciously used the
analytic approach in both his mathematical and his philosophical
investigations. In recent times, the method of working backwards
has been discussed by G. Polya.

The method can be relevant to any problem where we wish to
discover a process of a certain kind that leads from one point to
another. One example of such a process is the construction of a
derivation in an axiom system, but there are also many others. For
example, we may be looking at a map of a city and trying to
discover the best route from one point x to another point y.
Although our journey is to begin from x, it may in some cases be
easier to find the route by tracing it back from y. In general, when
seeking a path from one point to another, we often come to places
along the path where several alternative directions present themselves. The more such choices arise, the more difficult it is to find
the correct path. Working backwards can thus be useful when the
choices that we have to make in that manner are fewer or less
puzzling than those that we would need to make on a forward
journey. This is why a driver usually finds it easier to find his way
from a point on the edge of town to the town centre, rather than
vice versa.

Exercise 10   Imagine that we are presented with two formulae α
and β of truth-functional logic, and that we suspect that α does
not tautologically imply β. Describe two different ways of
searching for a verification of this suspicion, one of which could
be described as working forwards and the other as working
backwards. Illustrate the two different search procedures by
considering the example where α is $\neg p \vee (\neg p \vee r)$ whilst β is

$(\neg p \vee q) \vee r$ and also the example where α is $p \equiv (\neg q \vee r)$ whilst β is $\neg p \vee (q \equiv r)$.

Exercise 11　A man is beside a river with just two buckets. One bucket holds exactly nine litres and the other holds just four litres. He wants to bring up exactly two litres. How can he do it? Work backwards, beginning by imagining the task completed, and asking from what anterior result it could have been obtained, and so on.

Exercise 12　Work backwards once again to find a second solution to exercise 11.

Occasionally, working backwards enables us to avoid choices altogether. For example, children's puzzle books sometimes show a group of fishermen whose lines are entangled. On one of the hooks is a fish: whose is it? If we follow the lines downwards we have to make an initial choice between several alternatives, but if we follow the line upwards from the fish, no choice is required. On some other occasions, working backwards can lead to a more subtle result. It can enable us to reduce the choices from an infinity to some finite number, an upper bound upon whose size can be calculated in advance of any actual choosing. This situation almost, but not quite, arises in connection with the present axiom system; only the presence of the rule of transitivity prevents it from actually doing so. Axiom systems that do in fact satisfy this and related conditions are of considerable theoretical interest, and have been studied in depth by G. Gentzen. However we shall not pursue these questions any further here, but instead return to the derivation of theses in our axiom system.

So far we have examined three distribution principles. The first occurred as an axiom scheme, and the second and third were shown to be theses, without, incidentally, making any use of the first distribution principle in the derivations. The next exercise deals with the fourth distribution principle. It is considerably more difficult to derive than either of its predecessors, whether we work forwards or backwards. The derivation makes essential use of the axiom scheme of distribution.

Exercise 13   Let α, β and γ be any formulae. Show that the expression $(\alpha \vee \beta) \wedge (\alpha \vee \gamma) \rightarrow \alpha \vee (\beta \wedge \gamma)$ is a thesis of the axiom system.

Exercise 14   Let α and β be any formulae. Show that the following 'de Morgan principles' are theses of the axiom system:

$$\neg(\alpha \wedge \beta) \rightarrow \neg\alpha \vee \neg\beta \qquad \neg\alpha \vee \neg\beta \rightarrow \neg(\alpha \wedge \beta)$$
$$\neg(\alpha \vee \beta) \rightarrow \neg\alpha \wedge \neg\beta \qquad \neg\alpha \wedge \neg\beta \rightarrow \neg(\alpha \vee \beta).$$

Exercise 15   Let α and β be any formulae. Show that the following 'Lewis principles' are theses of the axiom system:

$$\alpha \wedge \neg\alpha \rightarrow \beta \qquad \alpha \rightarrow \beta \vee \neg\beta.$$

The de Morgan principles are so called after Augustus de Morgan, a logician of the nineteenth century. The Lewis principles are named after Clarence Irving Lewis who, in the early twentieth century, questioned their desirability. We shall return to the Lewis principles in the third section of this chapter.

Exercise 16   Show that the axiom scheme of disjunctive syllogism may be replaced by the Lewis principle $\alpha \wedge \neg\alpha \rightarrow \beta$ without altering the strength of the axiom system. Part of the answer is given by exercise 15.

2   Some tools for investigating systems of logic

If we are to study our axiom system further, we need some tools. As in carpentry, the tools may be simple or complex. In this section we shall provide a minimal kit for basic work on systems of logic. These tools are drawn from two main sources: arithmetic and set theory.

From arithmetic, the principal tool needed is the principle of proof by *induction over the positive integers*. This principle admits of several forms. Of them, the simplest and most basic can be put as follows. If the number one has a certain property, and if more-over $n + 1$ has the property whenever n has it, then every positive integer has the property. Thus if we wish to prove in arithmetic that every positive integer has a certain property, it suffices to show two things:

(1) That the number one has the property in question. This is called the *basis* of the induction;

(2) That for every positive integer n, if n has the property in question, then $n + 1$ also has it. This is called the *induction step* of the induction.

As an example, imagine that we want to show that every positive integer x has the property that $2^1 + \cdots + 2^x = 2^{x+1} - 2$. By the principle of induction it suffices to show two things:

(1) That the number one has the property – in other words that $2^1 = 2^{1+1} - 2$;

(2) That for every positive integer n, if n has the property then $n + 1$ has it too – in other words that if $2^1 + \cdots + 2^n = 2^{n+1} - 2$ then $2^1 + \cdots + 2^{n+1} = 2^{(n+1)+1} - 2$.

Now the first of these two points is immediate, since $2^1 = 2 = 2^2 - 2$. As for the second point, it can be verified quite easily as follows. Let n be any positive integer and suppose that $2^1 + \cdots + 2^n = 2^{n+1} - 2$. This supposition is known as the *induction hypothesis*. We want to show that $2^1 + \cdots + 2^{n+1} = 2^{(n+1)+1} - 2$. Now,

$$2^1 + \cdots + 2^{n+1} = (2^1 + \cdots + 2^n) + 2^{n+1}$$

and so, applying the induction hypothesis to the material in parentheses,

$$
\begin{aligned}
&= (2^{n+1} - 2) + 2^{n+1} \\
&= 2^{n+1} + 2^{n+1} - 2 \\
&= (2 \times 2^{n+1}) - 2 \\
&= 2^{n+2} - 2
\end{aligned}
$$

which completes the proof. In this way a significant arithmetic theorem emerges as the result of two quite simple steps.

Exercise 17 Use the principle of induction over the positive integers to show that for every positive integer x, the sum of the first x odd integers is equal to x^2. In other words, show that for every positive integer x, $1 + 3 + 5 + \cdots + (2x - 1) = x^2$. Set the proof out carefully on the model of the example given, distinguishing between the basis and the induction step, and stating explicitly what is to be shown in each. Also identify explicitly the induction hypothesis.

Exercise 18 Use the principle of induction over the positive integers to show that, for every positive integer x, the sum of all the positive integers less than or equal to x is equal to $x(x + 1)/2$. Follow the same instructions as in the preceding exercise.

Exercise 19 Use the principle of induction over the positive integers to show that in any reception attended by x people, if everybody shakes hands just once with everybody else, there are $x(x - 1)/2$ handshakes. Follow the same instructions as in exercise 17.

There are also other and more complex forms of induction over the positive integers. One very useful form is known as *course of values induction*. This principle tells us that if the number one has a certain property, and if moreover $n + 1$ has the property whenever all positive integers less than or equal to n have the property, then every positive integer has the property. At first glance, course of values induction would appear to be an even stronger principle than plain induction, since it contains a weaker antecedent, but in fact it is deducible from plain induction within the context of the usual axiomatization of arithmetic. We shall not go into the details here, as the deduction is more properly considered as part of arithmetic; we shall merely treat course of values induction as a tool, available for use whenever required.

There are several ways in which the principle of induction over the positive integers may be used in the study of systems of logic. We mention here just one application, that will arise in the following section. Imagine that we have defined a sequence $X_1, X_2, X_3, \ldots$ of items of some kind, one item for each positive integer. Imagine that we want to show that all of these items have a certain property. By the principle of induction it suffices to show that the first item in the sequence has the property in question, and that whenever the nth item in the sequence has the property, then so too does the $(n + 1)$th item. If we wish to use course of values induction, then it is enough to show that the first item in the sequence has the property in question, and that whenever all the items in the sequence up to and including the nth have the property, then so too does the $(n + 1)$th item.

The remainder of our minimal tool kit comes from set theory. If X and Y are sets, then we write $X \cup Y$, called the *union* of X and Y, to indicate the set whose elements are just those things that are elements of at least one of the sets X and Y. We write $X \cap Y$, called the *intersection* of X and Y, to indicate the set whose elements are just those things that are both elements of X and elements of Y. We write ϕ to stand for the set that has no elements at all, known as the *empty set*. If x is any object, we write $\{x\}$, called the *singleton* of x, to indicate the set whose only element is x. If x and y are any objects, not necessarily distinct from each other, we write $\{x, y\}$, called the *pair* made up of x and y, to indicate the set whose only elements are x and y. Similarly, if $x_1, \ldots, x_n$ are any objects, not necessarily mutually distinct, we write $\{x_1, \ldots, x_n\}$ to stand for the set whose elements are just $x_1, \ldots, x_n$. If we have a sequence $X_1, X_2, X_3, \ldots$ of sets, one for each positive integer, then we write $\bigcup X_i (i \geq 1)$ to indicate the set whose elements are just those things that are elements of at least one of the sets $X_1, X_2, X_3, \ldots$. We call $\bigcup X_i (i \geq 1)$ the *union* of the sequence $X_1, X_2, X_3, \ldots$. Clearly this notion is a generalization of the concept of the union of just two sets. If X and Y are sets we say that they are *identical*, and write $X = Y$, iff they have exactly the same elements. We say that X is a *subset* of Y, and write $X \subseteq Y$, iff every element of X is an element of Y. We say that X is a *proper subset* of Y iff it is a subset of Y but not identical with Y.

Exercise 20 Let X be the set of all tautologies, let Y be the set of all countertautologies, let Z be the set of all contingent formulae, and let α be the formula $p \vee \neg p$. Identify the following sets: $X \cup Y$, $X \cap Y$, $\{\alpha\}$, $X \cup (Y \cup Z)$, $(X \cup Y) \cup Z$, $X \cap (Y \cup Z)$, $X \cup \{\alpha\}$, $X \cap \{\alpha\}$, $Y \cup \{\alpha\}$, $Y \cap \{\alpha\}$. Indicate any identities that you notice between these sets.

Exercise 21 Let X and Y be any sets. Show that $X \cap Y \subseteq X$, $X \cap Y \subseteq Y$, $X \subseteq X \cup Y$, $Y \subseteq X \cup Y$. Also show that the follow-

ing three conditions are mutually equivalent: $X \cap Y = X, X \subseteq Y,$
$X \cup Y = Y.$

Exercise 22   For each positive integer i, let A_i be the set of all
those formulae of truth-functional logic that contain exactly i
distinct propositional letters. Identify the following sets: $A_1, A_2,$
$A_1 \cup A_2, A_1 \cap A_2, \bigcup A_i (i \geqslant 1), \bigcup A_i (i \geqslant 5).$

Exercise 23   Let $X_1, X_2, X_3, \ldots$ be a sequence of sets, one for
each positive integer. Show that for each $n, X_n \subseteq \bigcup X_i (i \geqslant 1).$

Exercise 24   Let $\{x, y\}$ and $\{x', y'\}$ be pairs. Show that
$\{x, y\} = \{x', y'\}$ iff either $x = x'$ and $y = y'$, or $x = y'$ and $y = x'.$

Another concept that we shall need from set theory is that of an
ordered pair $\langle x, y \rangle$, and more generally, an *ordered n-tuple*
$\langle x_1, \ldots, x_n \rangle$. Ordered pairs are not the same as ordinary pairs. If
$\langle x, y \rangle$ and $\langle x', y' \rangle$ are ordered pairs, then we say that $\langle x, y \rangle = \langle x', y' \rangle$
iff $x = x'$ and $y = y'$. This contrasts with the situation for ordinary
pairs, as can be seen from the result of exercise 24. More generally,
if $\langle x_1, \ldots, x_n \rangle$ and $\langle x'_1, \ldots, x'_n \rangle$ are ordered n-tuples, where n is
any positive integer, then we say that $\langle x_1, \ldots, x_n \rangle = \langle x'_1, \ldots, x'_n \rangle$
iff $x_1 = x'_1$ and $x_2 = x'_2$ and .... and $x_n = x'_n$; whereas in the case
of unordered n-tuples, we have $\{x_1, \ldots, x_n\} = \{x'_1, \ldots, x'_n\}$ iff
every element of the left set is an element of the right set, and every
element of the right set is an element of the left set, in any order
at all.

It turns out that the concept of an ordered pair can be defined
in terms of that of an ordinary, unordered, pair. There are in fact
several different ways of doing this. One of the most natural
procedures, though not the one most commonly used, is to select
two distinct objects a and b that are external to whatever domain
of discourse is under consideration, and to define $\langle x, y \rangle$ as
$\{\{a, x\}, \{b, y\}\}$. Another procedure, less natural to the mind but
formally more elegant, is to define $\langle x, y \rangle$ as $\{\{x\}, \{x, y\}\}$. Once
ordered pairs have been defined, ordered triples can be defined by
putting $\langle x_1, x_2, x_3 \rangle$ to be $\langle\langle x_1, x_2 \rangle, x_3 \rangle$, and so on. Whichever of
these two definitions of an ordered pair is followed, it yields the
basic property already mentioned, that for any objects $x, x', y, y',$

$\langle x, y \rangle = \langle x', y' \rangle$ iff $x = x'$ and $y = y'$. If desired, this result can easily be verified as an exercise.

For a first example of the application of concepts of set theory to the study of our axiom system, we return to the definition of a thesis. We have defined a thesis of the axiom system to be any expression of the form $\alpha \to \beta$, where α and β are formulae, that is either an instance of an axiom scheme or else is derivable from instances of the axiom schemes by means of a finite number of applications of the derivation rules. Now this definition is equivalent to another one. Let us say that a set X of expressions is *broad* iff it satisfies the following two conditions:

(1) Every instance of each of the axiom schemes is an element of X;

(2) Whenever we apply a derivation rule to expressions that are elements of X, then the expression obtained is an element of X.

Now the set of all theses of the axiom system is clearly a broad set, but it is not the only broad set. For example, the set of all expressions $\alpha \to \beta$ whatsoever, where α and β are any formulae, is also broad. We want to mark off the set of all theses as the *smallest* among the broad sets. To this end, we say that an expression is a thesis of the axiom system iff it is an element of *every broad set* of expressions.

This definition is rather more complex than its predecessor, but not perversely so. It serves a very useful purpose, for it leads us to a simple technique of proof. If we want to prove that every thesis of our axiom system has a certain property, it is sufficient to prove two things:

(1) Every instance of each of the axiom schemes has the property in question;

(2) Whenever a derivation rule is applied to expressions that have the property in question, then the expression obtained also has the property in question.

Why do these suffice? Let X be the set of all expressions that have the property in question. Suppose that the above two conditions are true. Then X satisfies the two clauses of the definition of a broad set, and so it is broad. Thus by the definition of the set of all theses, every thesis is an element of *X*, which is to say that every thesis has the property in question. We shall exploit this technique in the following section.

Clearly there is a resemblance between this technique of proof and the method of induction over the positive integers, described earlier in this section. For this reason, this kind of proof is also said to be *inductive*, and so too is the set-theoretic definition on which it rests. Within the proof, we can distinguish a basis and an induction step, as in the case of induction over the positive integers. The *basis* lies in showing that every instance of the axiom schemes has the property in question, and the *induction step* consists in proving that whenever a derivation rule is applied to expressions that have the property in question, then the expression obtained also has that property.

A deeper study of arithmetic and of set theory would probe further into the connections between these two kinds of inductive proof. It turns out that induction over the positive integers can be assimilated, as a special case, into the general pattern of set-theoretic induction. However we shall leave these questions aside, and merely use the two kinds of induction as tools for the study of systems of logic.

Exercise 25 Show that the original definition of the set of all theses of the axiom system is equivalent to the new definition. To do this, let *A* be the set of all theses in the original sense, and let *B* be the set of all theses in the new sense; show separately the two inclusions $B \subseteq A$ and $A \subseteq B$.

Exercise 26 Give two definitions of the set of all formulae of truth-functional logic, parallel in their style to the two definitions that have been given of the set of all theses of the axiomatic system. Indicate how each definition can be used in proving that all the formulae of truth-functional logic have a certain property.

The material of this section was introduced for a very practical purpose, but it also has a certain philosophical significance. In order to define the concept of a thesis of an axiomatization of the most elementary part of logic, or for that matter even to define the very notion of a formula of that part of logic, it is convenient to use concepts of set theory or of arithmetic. Even our original definition of the set of all theses makes implicit appeal to ideas of arithmetic with its reference to a finite number of applications of the derivation rules. Quite generally, we can say that the definition and study of any formal system at all, whether it be of logic or of anything else, is most conveniently carried out using not only notions of truth-functional and quantificational logic (for in our definitions we have made free use of locutions such as 'if . . . then . . . ', 'either . . . or . . . ', 'every', and 'there is') but also notions from set theory or from arithmetic, that lie beyond or in some sense 'above' logic.

3 Soundness and completeness

We want now to show that the axiom system lives up to its promise. There are two points to be established.

(1) We need to show that whenever $\alpha \to \beta$ is a thesis of the axiom system, then α tautologically implies β. This result can be expressed by saying that the axiom system is *sound*, or in the terminology of some writers, consistent with respect to the relation of tautological implication.

(2) Conversely, we must also show that whenever α tautologically implies β then the expression $\alpha \to \beta$ is a thesis of the axiom system. This can be expressed by saying that the axiom system is *complete* with respect to the relation of tautological implication.

Soundness is easy to prove, whereas completeness is more difficult. We begin with the easier.

Theorem Let α and β be formulae of truth-functional logic, constructed using the operators $\neg$, $\wedge$, $\vee$. If the expression $\alpha \to \beta$ is a thesis of the axiom system then α tautologically implies β. In brief, the axiom system is sound with respect to the relation of tautological implication.

Proof We want to show in effect that every thesis of the axiom system has a certain property. Thus in view of the discussion in the preceding section, it is enough to prove two things:

(1) Whenever $\alpha \rightarrow \beta$ is an instance of an axiom scheme then α tautologically implies β;

(2) Whenever we apply a derivation rule to expressions that correspond to tautological implications, then the expression obtained also corresponds to a tautological implication.

Moreover, these two points are quite easy to verify, by considering in turn each of the axiom schemes and each of the derivation rules.

Exercise 27 Carry out in detail the verifications needed for the derivation rules.

We come now to the more difficult proof of completeness. Unfortunately, it does not seem possible to establish this result by a straightforward inductive argument, as in the case of soundness. The reason for this is itself of some interest. If we want to show that one class A is included in another class B, then an inductive argument is often appropriate if the former class, A, is defined inductively. Otherwise, a different strategy is usually needed. Now in the present situation, the class of all theses is defined inductively, but the class of all tautological implications is not. Induction is thus convenient for proving the inclusion of the former in the latter, but it is not sufficient for the converse.

It turns out to be convenient to approach the completeness theorem indirectly, via its contrapositive. We shall prove that whenever $\alpha \rightarrow \beta$ is not a thesis of the axiom system then α does not tautologically imply β. Now the most straightforward way of showing that one formula does not tautologically imply another is by constructing an assignment of truth values that makes the former come out true and the latter come out false. Our problem in proving the completeness theorem can thus be reduced to one of constructing a suitable assignment of truth values, when the materials at hand for the construction are the formulae α and β themselves, together with the supposition that the expression $\alpha \rightarrow \beta$ is not a thesis of the axiom system.

This describes the initial strategy of the proof, but before entering into its details we need a definition. Let X be any set of formulae and let ψ be any formula. We say that X *axiomatically yields* ψ iff there are formulae $\varphi_1, \ldots, \varphi_n$ (where $n \geqslant 1$) that are elements of X such that the expression $(\varphi_1 \wedge (\varphi_2 \wedge \ldots \wedge \varphi_n) \rightarrow \psi$ is a thesis of the axiom system. This concept is an abstraction upon the concept of thesishood, as the following exercises make clear.

Exercise 28   Show that the bracketing of the formulae $\varphi_2, \ldots, \varphi_n$ as also their order, is immaterial in the above definition.

Exercise 29   Show that a singleton $\{\varphi\}$ axiomatically yields a formula ψ iff the expression $\varphi \rightarrow \psi$ is a thesis of the axiom system.

Exercise 30   Show that every set of formulae axiomatically yields each of its elements.

Exercise 31   Let X and Y be sets of formulae and let ψ be a formula. Show that if X axiomatically yields ψ and X is a subset of Y, then Y axiomatically yields ψ.

Exercise 32   Let X be any finite set of formulae and let ψ be a formula. Show that X axiomatically yields ψ iff $\varphi \rightarrow \psi$ is a thesis, where φ is the conjunction, in some order, of *all* the formulae in X.

Exercise 33   Let X be any infinite set of formulae and let ψ be a formula. Show that X axiomatically yields ψ iff there is some finite subset of X that axiomatically yields ψ.

Theorem   Let α and β be formulae of truth-functional logic, constructed using the operators $\neg, \wedge, \vee$. If α tautologically implies β, then the expression $\alpha \rightarrow \beta$ is a thesis of the axiom system. In brief, the axiom system is complete with respect to the relation of tautological implication.

Proof   Suppose that $\alpha \rightarrow \beta$ is not a thesis of the axiom system. We want to show that α does not tautologically imply β.

Let $\gamma_1, \gamma_2, \gamma_3, \ldots$ be a list of *all* formulae of truth-functional logic that can be built by means of the operators $\neg, \wedge, \vee$. Clearly the list is infinite. We now define a sequence $A_1, A_2, A_3, \ldots$ of *sets* of formulae, as follows:

(1) The set A_1 is defined to be the singleton $\{\alpha\}$;

(2) Let i be any integer. We define the set A_{i+1} from the set A_i by the following rule:

      (a) If the set $A_i \cup \{\gamma_i\}$ axiomatically yields β, then the set A_{i+1} is defined to be just the set A_i;

      (b) If on the other hand the set $A_i \cup \{\gamma_i\}$ does not axiomatically yield β, then the set A_{i+1} is defined to be $A_i \cup \{\gamma_i\}$.

The definition of the sequence $A_1, A_2, A_3, \ldots$ is inductive, and its inductive step is disjunctive. For this reason, it may at first appear rather strange. However its general drift can be made clear by comparing it with an analogous physical process. Of course, the comparison itself does not form part of the proof.

Imagine that we have a box full of small objects like paper clips, pieces of chalk, pencil stubs, matches and drawing pins. We choose two of these objects and call them α and β. We take a balance, and put α in the left-hand pan, and β in the right-hand pan. Imagine that α *is not as heavy as* β, in other words, that the balance tilts to the right. This is analogous to our supposition that the expression $\alpha \rightarrow \beta$ is not a thesis of the axiom system. We now arrange all the objects in the box into a list, calling them $\gamma_1, \gamma_2, \gamma_3, \ldots$ and so on, and allocate them according to the following plan. Begin by taking the first object γ_1 and test it out by adding it to α in the left-hand pan. If the left-hand pan becomes as heavy as the right-hand pan, so that it reaches a horizontal position or tips leftwards, then we take γ_1 out again and throw it away. On the other hand, if when we test γ_1 the balance does not reach a horizontal position or tip leftwards, then we leave γ_1 in the left-hand pan. Now take the second object γ_2 and test it and allocate it in the same way. Then take the third object γ_3, and so on with all the others. In this way we slowly form a collection of objects in the left-hand pan. The same basic process can also be described by means of a flow chart, as follows.

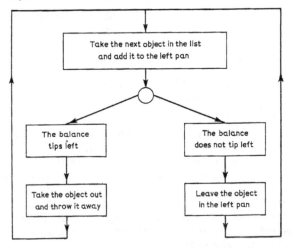

We now return to the proof itself. We define the set A to be the union $\bigcup A_i (i \geqslant 1)$ of the entire sequence $A_1, A_2, A_3, \ldots$ of sets of formulae. We claim that the set A has some very attractive and useful properties.

Claim 1   The formula α is an element of A.

Verification   By definition, α is an element of A_1, and so it is an element of A.

Claim 2   For each integer i, the set A_i does not axiomatically yield β.

Verification   To show this, we argue inductively. We need to show first that A_1 does not axiomatically yield β, and second that for every integer i, if A_i does not axiomatically yield β, then A_{i+1} does not axiomatically yield β. For the first, we observe that $A_1 = \{\alpha\}$, and recall that by hypothesis the expression $\alpha \rightarrow \beta$ is not a thesis of the axiom system. As for the second, it is immediate from the definition of A_{i+1}.

Claim 3   The set A does not axiomatically yield β.

Verification   Suppose for *reductio ad absurdum* that A does axiomatically yield β. Then there are formulae $\varphi_1, \ldots, \varphi_n$ in A such that the expression $\varphi_1 \wedge \cdots \wedge \varphi_n \to \beta$ is a thesis of the axiom system. Now since each of $\varphi_1, \ldots, \varphi_n$ is an element of A, each of $\varphi_1, \ldots, \varphi_n$ is an element of some set in the list $A_1, A_2, A_3, \ldots$. Hence since A_i is always a subset of A_{i+1}, there is a set A_k in the list such that each of $\varphi_1, \ldots, \varphi_n$ is an element of A_k. But this means that A_k axiomatically yields β, which contradicts claim 2.

Claim 4   The formula β is not an element of A.

Verification   If β is an element of A, then by exercise 30, A axiomatically yields β, which contradicts claim 3.

Claim 5   For every formula ψ, if A axiomatically yields ψ then ψ is an element of A.

Verification   Let ψ be a formula, and suppose for *reductio ad absurdum* that A axiomatically yields ψ but ψ is not an element of A. Now since A axiomatically yields ψ, there are formulae $\varphi_1, \ldots, \varphi_n$ in A such that the expression

$$\varphi_1 \wedge \cdots \wedge \varphi_n \to \psi$$

is a thesis of the axiom system. Also, since the list $\gamma_1, \gamma_2, \ldots$ covers *all* formulae, we can say that ψ is the formula γ_i for some $i \geqslant 1$. Further, since γ_i is not an element of A, it is not an element of A_{i+1}. Hence by the definition of A_{i+1}, the set $A_i \cup \{\gamma_i\}$ axiomatically yields β. Thus there are formulae $\theta_1, \ldots, \theta_m$ in A_i such that the expression

$$\theta_1 \wedge \cdots \wedge \theta_m \wedge \gamma_i \to \beta$$

is a thesis of the axiom system. Putting the two displayed results together, it follows that the expression

$$\theta_1 \wedge \cdots \wedge \theta_m \wedge \varphi_1 \wedge \cdots \wedge \varphi_n \to \beta$$

is a thesis of the axiom system. But $\theta_1, \ldots, \theta_m$ are all elements of A_i, and so of A, and $\varphi_1, \ldots, \varphi_n$ are also elements of A. Thus A axiomatically yields β, which contradicts claim 3.

Claim 6   The set A is 'well behaved' with respect to conjunction, in the following sense: for all formulae φ and ψ, $\varphi \wedge \psi$ is an element of A iff both φ and ψ are elements of A.

Verification   This can easily be deduced from claim 5.

Claim 7   The set A is 'well behaved' with respect to disjunction, in the following sense: for all formulae φ and ψ, $\varphi \vee \psi$ is an element of A iff at least one of φ and ψ is an element of A.

Verification   One half of this can easily be deduced from claim 5. As for the other half, suppose for *reductio ad absurdum* that $\varphi \vee \psi$ is an element of A but neither φ nor ψ is an element of A. Now since the list $\gamma_1, \gamma_2, \ldots$ covers *all* formulae, $\varphi = \gamma_i$ and $\psi = \gamma_j$ for some integers i and j. Since γ_i is not an element of A, it is not an element of A_{i+1}. Hence by the definition of A_{i+1}, the set $A_i \cup \{\gamma_i\}$ axiomatically yields β. Thus there are formulae $\theta_1, \ldots, \theta_n$ in A_i such that the expression

$$\theta_1 \wedge \ldots \wedge \theta_n \wedge \gamma_i \rightarrow \beta$$

is a thesis. Using claim 6 we can say that there is a formula θ in A such that the expression

$$\theta \wedge \gamma_i \rightarrow \beta$$

is a thesis. A similar argument applied to γ_j tells us that there is a formula θ' in A such that the expression

$$\theta' \wedge \gamma_j \rightarrow \beta$$

is a thesis. Hence clearly the expressions

$$\theta \wedge \theta' \wedge \gamma_i \rightarrow \beta$$
$$\theta \wedge \theta' \wedge \gamma_j \rightarrow \beta$$

are both theses, and so using the rule for disjunction of antecedents, the expression

$$(\theta \wedge \theta' \wedge \gamma_i) \vee (\theta \wedge \theta' \wedge \gamma_j) \rightarrow \beta$$

is a thesis. So by the axiom scheme of distribution, together with the rule of transitivity, the expression

$$(\theta \wedge \theta') \wedge (\gamma_i \vee \gamma_j) \rightarrow \beta$$

is a thesis. But we know that both θ and θ' are in A, and moreover by hypothesis the formula $\varphi \vee \psi$, which is $\gamma_i \vee \gamma_j$, is an element of A. Thus A axiomatically yields β, which contradicts claim 3.

Claim 8   The set A is 'well behaved' with respect to negation, in the following sense: for every formula φ, $\neg\varphi$ is an element of A iff φ is not an element of A.

Verification   We need to show, in effect, that for any formula φ, one and only one of φ and $\neg\varphi$ is an element of A. To show one half of the claim, suppose for *reductio ad absurdum* that both φ and $\neg\varphi$ are elements of A. Now we know from exercise 15 that $\varphi \wedge \neg\varphi \rightarrow \beta$ is a thesis of the axiom system, so A axiomatically yields β, which contradicts claim 3. It remains to establish the other half of the claim, that either φ is an element of A or $\neg\varphi$ is an element of A. Recall from exercise 15 that $\alpha \rightarrow \varphi \vee \neg\varphi$ is a thesis of the axiom system, and recall from claim 1 that α is an element of A. Thus A axiomatically yields $\varphi \vee \neg\varphi$ and so by claim 5, $\varphi \vee \neg\varphi$ is an element of A, and so by claim 7, at least one of φ and $\neg\varphi$ is an element of A.

These eight claims bring us to the final step of the proof. We define an assignment v of truth values to propositional letters as follows. For each propositional letter p, we assign p the value 'true' if p is an element of A, and we assign p the value 'false' if p is not an element of A. Now, using the fact that the operators of conjunction, disjunction, and negation are all 'well behaved' in the senses specified in claims 6, 7 and 8, we can easily verify that for *every* formula φ whatsoever, if φ is an element of A then φ comes out true under the assignment v, whilst if φ is not an element of A then φ comes out false under the assignment v.

But by claims 1 and 4, α is an element of A whilst β is not. Thus α comes out true under the assignment v, whilst β comes out false under v, so that α does not tautologically imply β. This completes the proof.

Exercise 34   Carry out in detail the verification of claim 6.

Exercise 35   Give the details of the verification of the first half of claim 7.

Exercise 36   Carry out in detail the verification involved in the penultimate paragraph of the proof, using an inductive argument. To make the exposition brief, write $v(\varphi) = 1$ as an abbreviation

for 'φ comes out true under the assignment v' and write $v(\varphi) = 0$
for 'φ comes out false under the assignment v'.

There are many other ways of proving that our axiom system is complete with respect to the relation of tautological implication, and also many ways of rearranging the material in the present proof. However the present argument has some important advantages. It can be extended, without much trouble, to obtain an analogous theorem in quantificational logic. It can also be modified without much difficulty to obtain certain completeness results for some nonclassical logics.

In this connection, we should note for future reference the way in which the proof handles the axiom scheme of disjunctive syllogism. That scheme was used in establishing claim 8 of the proof: in effect, we observed there that formulae of the forms $\varphi \wedge \neg \varphi \rightarrow \beta$ and $\alpha \rightarrow \varphi \vee \neg \varphi$ are theses of the axiom system, and that result was established in exercise 15 with the help of the axiom scheme of disjunctive syllogism. But nowhere before claim 8 does the proof use the fact that disjunctive syllogism is an axiom scheme of the system. This means that the entire proof, up to the end of claim 7, is also valid for the axiom system like the present one but lacking disjunctive syllogism. We shall return to this point in the next chapter.

2 Some modified implication relations

Outline

We begin by pointing out some of the eccentricities of the relation of tautological implication, and formulating the problem of entailment. We then describe some modified implication relations that have been suggested in connection with this problem. Two of these relations, subtending and tautopical implication, are discussed briefly. A third, de Morgan implication, is studied in some depth. In particular, it is shown how de Morgan implication can be characterized as a four valued logic, and how these four values may be associated with the four subsets of the set of the two classical truth values.

1 The problem of entailment

The first chapter was centred upon a formal problem, setting out an axiom system for the relation of tautological implication and proving its soundness and completeness. In this chapter we shall look at some more informal questions, concerning the oddities of tautological implication and some ideas for avoiding them.

By and large, tautological implication is an unsurprising relation. It is no surprise, for example, to learn that a conjunction implies each of its conjuncts or that a disjunction is implied by each of its disjuncts. Even the distribution principles turn out to be more or less what we would expect. But there are some examples of tautological implication that can appear strange, and in particular the Lewis principles $\alpha \wedge \neg\alpha \to \beta$ and $\alpha \to \beta \vee \neg\beta$ mentioned in exercise 15.

Exercise 37 Use the definition of tautological implication to establish the following more general version of the Lewis principles: for all formulae α and β, α tautologically implies β whenever β is a tautology or α is a countertautology.

Different people react in different ways to the Lewis principles. For some they are welcome guests, whilst for others they are strange and suspect. For some, it is no more objectionable in logic to say that a countertautology implies all formulae than it is in arithmetic to say that x^0 always equals 1. In each case, we are merely extending a definition to cover a limiting case in the smoothest and most convenient possible way. For others, however, the Lewis principles are quite unacceptable because the antecedent formulae may have 'nothing to do with' the consequent formulae. For example, the statement

Mercury is both a metal and not a metal

has nothing to do with the statement

Shakespeare was a woman in disguise

and yet the former tautologically implies the latter. It has been suggested, along these lines, that in any logical implication worthy of the name, the antecedent should be *relevant* to the consequent. There should perhaps be some *connection between the meanings* of the statements filling the antecedent and the consequent positions. The relation should be in some sense *essentially relational*: no implication should hold in virtue of its antecedent alone, nor its consequent alone, but in virtue of a link between them.

Even if we are hesitant about such assertions, an intriguing problem remains: to inquire whether there are any interesting and reasonably systematic relations that avoid the Lewis principles, and satisfy conditions like those mentioned above more adequately than does the relation of tautological implication. This is known as *the problem of entailment*.

The first point to observe in connection with the problem of entailment is that any implication that gives up the Lewis principle $\alpha \wedge \neg\alpha \rightarrow \beta$ must *also* give up at least one among certain other principles that appear much more innocent and natural. For as we observed in exercise 15, the scheme $\alpha \wedge \neg\alpha \rightarrow \beta$ can be derived in our axiom system in the following way.

1. $\alpha \wedge \neg\alpha \rightarrow \alpha$ axiom scheme for conjunction
2. $\alpha \rightarrow \alpha \vee \beta$ axiom scheme for disjunction
3. $\alpha \wedge \neg\alpha \rightarrow \alpha \vee \beta$ 1 and 2, rule of transitivity
4. $\alpha \wedge \neg\alpha \rightarrow \neg\alpha$ axiom scheme for conjunction
5. $\alpha \wedge \neg\alpha \rightarrow (\alpha \vee \beta) \wedge \neg\alpha$ 3 and 4, rule for conjunction
6. $(\alpha \vee \beta) \wedge \neg\alpha \rightarrow \beta$ axiom scheme of disjunctive syllogism
7. $\alpha \wedge \neg\alpha \rightarrow \beta$ 5 and 6, rule of transitivity

Thus any relation between formulae that satisfies the axiom schemes for conjunction, disjunction, and disjunctive syllogism, together with the rules of transitivity and conjunction in the consequent, must also satisfy the Lewis principle $\alpha \wedge \neg\alpha \rightarrow \beta$. Put in other words, any relation that gives up the Lewis principle must give up at least one of these apparently innocuous axiom schemes and derivation rules. This observation is known as *Lewis's dilemma*.

To many, giving up any of these principles is much *more* unnatural than accepting the Lewis principles. Even so, however, the problem of entailment retains its meaning. Of the various ways of modifying the relation of tautological implication so as to avoid the Lewis principles, which are the more significant? We shall look at several ideas, some of more interest than others.

2   *Subtending and tautopical implication*

One idea begins by focusing attention on the difference between the expression $p \wedge \neg p \rightarrow q$, where p and q are distinct proposition letters, and the expression $p \wedge \neg p \rightarrow p$. In each case the antecedent is a countertautology, but the latter expression differs from the former in that it can be seen as a substitution instance of the expression $p \wedge q \rightarrow p$, whose antecedent is not a countertautology. In general, if α and β are formulae, we can say that α *subtends* β if the expression $\alpha \rightarrow \beta$ is a substitution instance of some expression $\alpha_0 \rightarrow \beta_0$ such that α_0 tautologically implies β_0, but where α_0 is not a countertautology and β_0 is not a tautology.

Clearly, this is a subrelation of tautological implication: whenever α subtends β then α tautologically implies β. Also, the relation permits many of the more familiar principles. For example, a conjunction subtends each conjunct, and a disjunction is subtended by each disjunct. Moreover, the relation avoids the Lewis principles: $p \wedge \neg p$ does not subtend q, and p does not subtend $q \vee \neg q$, where p and q are distinct propositional letters.

> *Exercise 38* Check in detail the last assertion made, by examining the possible kinds of expression of which $p \wedge \neg p \rightarrow q$ and $p \rightarrow q \vee \neg q$ could be substitution instances.

However, subtending pays a price for avoiding the Lewis principles. It has at least four awkward features.

(1) The scope of the relation depends upon the choice of the primitive operators. To explain this remark, recall that in this book we are working with the primitive operators $\neg$, $\wedge$, $\vee$ and that the operators $\supset$ and $\equiv$ are introduced as abbreviating devices. In such a situation, the formula $p \equiv q$ subtends the formula $p \equiv p$. For the expression $p \equiv q \rightarrow p \equiv p$ is an abbreviation for the more complex expression

$$\neg(p \wedge \neg q) \wedge \neg(q \wedge \neg p) \rightarrow \neg(p \wedge \neg p) \wedge \neg(p \wedge \neg p)$$

and this expression is a substitution instance of a tautological implication whose antecedent is not a countertautology and whose consequent is not a tautology, namely

$$\neg(p \wedge \neg q) \wedge \neg(q \wedge \neg r) \rightarrow \neg(p \wedge \neg r) \wedge \neg(p \wedge \neg r).$$

The former expression can be obtained from the latter by substituting p for r. On the other hand, if we had included $\equiv$ among our primitive operators, then the situation would have been quite different. In such a case, $p \equiv q$ would not subtend $p \equiv p$, because there is no appropriate expression of which $p \equiv q \rightarrow p \equiv p$ is a substitution instance.

> *Exercise 39* Check in detail the last assertion made, by examining the possible kinds of expression of which $p \equiv q \rightarrow p \equiv p$ could be a substitution instance.

(2) The relation of subtending is not transitive. For example, $p \wedge \neg p$ subtends $(p \vee q) \wedge \neg p$, and the latter subtends q, but as we have already seen, $p \wedge \neg p$ does not subtend q.

Exercise 40   Verify that $p \wedge \neg p$ subtends $(p \vee q) \wedge \neg p$, and that the latter subtends q.

(3) Subtending does not satisfy the rule of conjunction in the consequent. For example, $(p \vee q) \wedge (\neg p \wedge \neg q)$ subtends each of q and p considered separately, but it does not subtend their conjunction.

Exercise 41   Verify the above remarks.

(4) Subtending does not satisfy the rule of disjunction in the antecedent.

Exercise 42   Give an example that illustrates the above remark. Choose the example in the light of the preceding remark.

For these reasons, subtending seems to be a rather unattractive relation, and hardly suitable as an alternative to tautological implication. Another relation, that seems to pay a lower price, is the following. We define topic-restricted tautological implication, or more briefly *tautopical implication*, to be the relation that holds from α to β iff α tautologically implies β and moreover every propositional letter occurring in the consequent β already occurs in the antecedent α.

This relation avoids the Lewis principles. The formula $p \wedge \neg p$ does not tautopically imply the formula q, because the propositional letter q occurs in the consequent but not in the antecedent. For the same reason, p does not tautopically imply $q \vee \neg q$. It is easy to show that the relation is transitive, and that it satisfies the rules of conjunction in the consequent and disjunction in the antecedent. Of course, we know from Lewis's dilemma that it must give up something somewhere, and its losses are at least two.

(1) The relation does not satisfy the axiom scheme for disjunction. In particular, if p and q are distinct propositional letters, then p does not tautopically imply $p \vee q$.

(2) The relation does not satisfy the rule of contraposition. For example, $p \wedge q$ tautopically implies p, but $\neg p$ does not tautopically imply $\neg (p \wedge q)$.

However all the remaining axiom schemes and derivation rules from our axiom system hold true for tautopical implication. It is thus a less radical departure from tautological implication pure and simple than its predecessor.

3   De Morgan implication

For a third approach to the problem of entailment, let us define *de Morgan implication* to be the relation between formulae that is determined by taking our axiom system for tautological implication, and simply dropping from it the axiom scheme of disjunctive syllogism.

This relation comes quite close to tautological implication. By its definition, it satisfies all of the axiom schemes and derivation rules of our axiom system, except one. Moreover, the principles of identity, commutation, association, absorption, distribution and de Morgan are theses of the new system as well as of the old, for in their derivations, carried out in exercises 5–9 and 13–14 in chapter I, we made no use of the axiom scheme of disjunctive syllogism. It is the presence of the de Morgan principles, as prominent theses of this kind of implication, that is the rationale for its name.

On the other hand, the standard derivation of the Lewis principle $p \wedge \neg p \to q$ now breaks down at its sixth step. Moreover, it turns out that it is not possible to bypass this obstacle by constructing a longer or more circuitous derivation. It can be shown that the expression $p \wedge \neg p \to q$ is not derivable at all in the system for de Morgan implication, but to prove this we need first to look at models. We shall show that de Morgan implication can be characterized as a four valued logic.

We choose the letters 1, 0, *a, b* to designate four distinct objects that will serve as values. It does not matter *what* the objects *a* and *b* are, although we shall describe an interesting way of interpreting them later. It is tempting to think of *a* and *b* as fractional numbers between 0 and 1, but this is *definitely not* a good policy, for it suggests a linear relationship between the four values, and as we shall soon see, the relationship is more complex than that: the values of *a* and *b* turn out to be in a certain sense 'incomparable'. We stipulate that the value of a negation, a conjunction, or a disjunction is determined from the values of its parts, in the way described in the following tables.

It should be clear how to read the table for negation. Those for conjunction and disjunction are meant to be read as follows. If α has value x and β has value y, then we look along the row from x and down the column under y to find the appropriate entry, which tells us the corresponding value of the conjunction, or disjunction as the

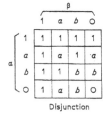

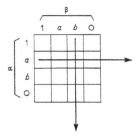

case may be. For example, suppose that $v(\alpha) = a$ and $v(\beta) = b$, and that we want to find the value of $v(\alpha \wedge \beta)$. We take the table for conjunction, and look along the row for a and down the column for b, as follows,

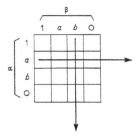

to discover, at the point of intersection, the value of $v(\alpha \wedge \beta)$ to be 0.

Exercise 43   If $v(\alpha) = a$ and $v(\beta) = b$, then what is the value of $v(\alpha \vee \beta)$?

Exercise 44   If $v(\alpha) = a$ and $v(\beta) = 1$, then what are the values of $v(\alpha \wedge \beta)$ and $v(\alpha \vee \beta)$?

The information contained in these tables can be set out much more strikingly in a diagram, called a *Hasse diagram*, and borrowed from lattice theory. First, let us focus attention on conjunction and disjunction, and represent them by the following diagram:

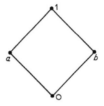

In this diagram, conjunctions are read downwards and disjunctions are read upwards. Thus if $v(\alpha) = a$ and $v(\beta) = b$ then $v(\alpha \wedge \beta) = 0$ and $v(\alpha \vee \beta) = 1$. Again, if $v(\alpha) = a$ and $v(\beta) = 1$, then $v(\alpha \wedge \beta) = a$ and $v(\alpha \vee \beta) = 1$. To complete the diagram, we add arrows to show the behaviour of negation:

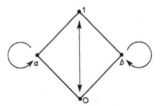

Thus if $v(\alpha) = 1$ then $v(\neg \alpha) = 0$, found by following the arrow from 1 to 0. If $v(\alpha) = b$ then $v(\neg \alpha) = b$, found by following the arrow from b to b.

We shall call this structure *the four element model*. Of course, one can construct many other models with just four elements, by changing the pattern of negation, say, or the pattern of one of the other operators. But when we refer to *the* four element model, we shall always mean this particular one. If x and y are elements of the model, we shall say that x is *less than or equal to* y, and write $x \leqslant y$, iff either x is the same as y, or else x occurs lower than y in the diagram. Thus for example, $a \leqslant a$ and $b \leqslant 1$, but $a \not\leqslant 0$ and $1 \not\leqslant b$. Observe that the elements a and b of the model are *incomparable*, in the sense that neither of them is less than or equal to the other: $a \not\leqslant b$ and $b \not\leqslant a$.

Thus the relation ≤ between elements of the model is not a linear one.

If α and β are formulae built up with the operators $\neg$, $\wedge$, $\vee$, we say that the expression $\alpha \to \beta$ is *valid* in the four element model iff for every assignment of values in the four element model to propositional letters, $v(\alpha) \leq v(\beta)$. Thus for example, the expression $\alpha \wedge \beta \to \alpha$ is valid, for any formulae α and β. This can be verified roughly by just looking at the diagram: for any assignment v of values in the model to propositional letters, $v(\alpha \wedge \beta)$ will be the same as, or else lower down in the diagram than, $v(\alpha)$. For a rigorous check we can proceed as follows. Let v be any assignment of values in the model to propositional letters. Since the model contains only four elements, there are just sixteen possible combinations of values for α and β. Check these one by one, using the table for conjunction: in each case it turns out that $v(\alpha \wedge \beta) \leq v(\alpha)$.

Exercise 45 Show, both in sketch and in detail, that any expression of the form $\alpha \to \alpha \vee \beta$ is valid in the four element model.

On the other hand, if p and q are distinct propositional letters, then the expression $(p \vee q) \wedge \neg p \to q$ is not valid in the four element model. For let v be the assignment that puts $v(p) = a$ and $v(q) = 0$. Then clearly $v((p \vee q) \wedge \neg p) = a$ and $v(q) = 0$, so that $v((p \vee q) \wedge \neg p) \not\leq v(q)$.

Exercise 46 Show that neither of the expressions $p \wedge \neg p \to q$ and $p \to q \vee \neg q$ is valid in the four element model.

Exercise 47 Show that the expression $p \wedge \neg p \to q \vee \neg q$ is not valid in the four element model.

Exercise 48 Let α and β be any formulae. Show that if α and β have no propositional letters in common, then the expression $\alpha \to \beta$ is not valid in the four element model. This result generalizes those of the preceding two exercises.

Theorem Let α and β be formulae built with the operators $\neg$, $\wedge$, $\vee$. If the expression $\alpha \to \beta$ is a thesis of the axiom system for de Morgan implication, then $\alpha \to \beta$ is valid in the four element model. In other words, the axiom system for de Morgan implication is sound with respect to validity in the four element model.

Proof To prove the theorem, it is enough to show two points:

(1) Whenever $\alpha \to \beta$ is an instance of an axiom scheme of the system for de Morgan implication, then $\alpha \to \beta$ is valid in the four element model;

(2) Whenever we apply a derivation rule of the system for de Morgan implication to expressions that are valid in the four element model, then the expression obtained is also valid in the four element model.

These two points are quite easy to verify, by considering each axiom scheme and each derivation rule in turn.

Exercise 49 Carry out in detail the verification needed for the derivation rule of contraposition.

We can use the theorem to establish negative results about de Morgan implication, and in particular to show that de Morgan implication does not admit the Lewis principles.

First corollary If p and q are distinct propositional letters, then neither of the expressions $p \wedge \neg p \to q$ and $p \to q \vee \neg q$ is a thesis of the axiom system of de Morgan implication. Nor is the expression $p \wedge \neg p \to q \vee \neg q$ a thesis of de Morgan implication.

Proof We have already observed in exercises 46 and 47 that none of these expressions is valid in the four element model. Hence by the theorem, none of them is a thesis of the axiom system for de Morgan implication.

More generally, we can also show that de Morgan implication satisfies a certain 'relevance condition'.

Second corollary Whenever an expression $\alpha \to \beta$ is a thesis of the system of de Morgan implication, then α and β share at least one propositional letter.

Proof We have already observed in exercise 48 that if α and β do not share any propositional letters, then the expression $\alpha \to \beta$ is not valid in the four element model, and so, by the soundness theorem, it is not a thesis of de Morgan implication.

Theorem Let α and β be formulae built using the operators $\neg$, $\wedge$,

V. If the expression $\alpha \to \beta$ is valid in the four element model, then $\alpha \to \beta$ is a thesis of the axiom system for de Morgan implication. In other words, the axiom system for de Morgan implication is complete with respect to validity in the four element model.

Proof To establish this result, we need only make a few adjustments to the proof of the corresponding completeness theorem for classical truth-functional logic. As before, we proceed indirectly, by proving the contrapositive. We suppose that $\alpha \to \beta$ is not a thesis of de Morgan implication, and we want to construct an assignment v into the four element model such that $v(\alpha) \not\leqslant v(\beta)$. To do this, we define as before a sequence $A_1, A_2, A_3, \ldots$ of sets of formulae and form the union $\bigcup A_i (i \geqslant 1)$ of the entire list. We then argue exactly as before, up to the end of claim 7. This is possible because, as we have observed, up to that point no use was made of the axiom scheme of disjunctive syllogism. We then proceed as follows.

Modified Claim 8 The set A is 'well behaved' with respect to negation in the following modest sense: for every formula φ, φ is an element of A iff $\neg\neg\varphi$ is an element of A.

Verification Suppose first that φ is an element of A. Now $\varphi \to \neg\neg\varphi$ is a thesis, and so A axiomatically yields $\neg\neg\varphi$. So by claim 5, $\neg\neg\varphi$ is an element of A. A similar argument can be used for the converse.

Now we define an assignment v of values in the four element model. For each propositional letter p we say,

(1) If p is an element of A and $\neg p$ is not an element of A, then put $v(p) = 1$;
(2) If p is not an element of A and $\neg p$ is an element of A, then put $v(p) = 0$;
(3) If p and $\neg p$ are both elements of A, then put $v(p) = a$;
(4) If neither p nor $\neg p$ is an element of A, then put $v(p) = b$.

Then we can verify, using claims 5, 6, 7, and the modified claim 8, that *every* formula φ satisfies conditions (1) through (4). Hence, since α is an element of A, either $v(\alpha) = 1$ or $v(\alpha) = a$. And since β is not an element of A, either $v(\beta) = 0$ or $v(\beta) = b$. Thus $v(\alpha) \not\leqslant v(\beta)$ and so the expression $\alpha \to \beta$ is not valid in the four element model. This completes the proof of the completeness theorem.

Exercise 50 Carry out in detail the verifications that are needed to show that every formula satisfies conditions (1) through (4). To do this, you will need to use an inductive argument. Within the induction step you will need to consider three separate cases, corresponding to the operators ¬, ∧, ∨. Within each case you will need to distinguish further subcases. However each subcase is quite easy.

It is interesting to ask whether we can give any meaning to the four values of our model, and in particular to the two new values a and b. An intriguing, if rather fanciful, answer to this question can be given by turning to Aristotle and his famous three 'laws of thought'. Aristotle's laws of thought can be stated in many different ways, but the formulation that is useful to us here is the following:

(1) The law of identity: any two occurrences of a single statement have the same value;

(2) The law of noncontradiction: no statement is both true and false;

(3) The law of excluded third: every statement is either true or false.

These three laws can be taken as instructions for assigning truth values to propositional letters in classical logic. Each letter is assigned the value 'true' or the value 'false' (in accord with excluded third), but not both (in accord with noncontradiction), and distinct occurrences of a single propositional letter always receive the same value (in accord with identity).

Now from a formal point of view, de Morgan implication may be regarded as a logic that abandons both the law of noncontradiction and the law of excluded third, whilst retaining the law of identity. To assign a propositional letter the value 1 is to stipulate that it is true and not false. To assign it the value 0 is to assert that it is false and not true. But to assign it the value a is to stipulate that it is both true and false, and to assign it the value b is to treat it as neither true nor false. The matter may be expressed more succinctly if we speak of sets. Classically, there are two truth values 1 and 0. The elements of the four element model may naturally be regarded as *the four subsets* of the set of the two classical values. The diagram that we

have been using may be re-labelled as follows, where ϕ stands for the
empty set:

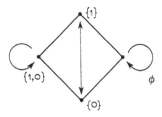

The interconnections between the values of the four element
model and these four subsets can be taken even further. For we can
easily check that for any formulae α and β, and any assignment v into
the four element model,

> 1 is in $v(\alpha \wedge \beta)$ iff 1 is in $v(\alpha)$ and also in $v(\beta)$
> 1 is in $v(\alpha \vee \beta)$ iff 1 is in $v(\alpha)$ or is in $v(\beta)$
> 1 is in $v(\neg\alpha)$ iff 0 is in $v(\alpha)$.

Dually, we also have

> 0 is in $v(\alpha \wedge \beta)$ iff 0 is in $v(\alpha)$ or is in $v(\beta)$
> 0 is in $v(\alpha \vee \beta)$ iff 0 is in $v(\alpha)$ and also in $v(\beta)$
> 0 is in $v(\neg\alpha)$ iff 1 is in $v(\alpha)$.

To sum up, in so far as its formal structure is concerned, de Morgan
implication can be treated as a nonclassical logic that departs from
Aristotle's laws of thought in a particularly simple and straightforward
way. Its values are just the four subsets of the set of the two classical
values, and the behaviour of these four is determined in a simple
manner by the behaviour of the classical values.

However it is rather difficult to see any intrinsic reason why we
should want to treat the propositions or statements of ordinary
language as capable of four values. Nor is it easy to envisage practical
criteria for determining which of these four values any given statement
should bear. For example, there seems to be little reason for regarding
the statement 'macaroni is fattening' as four valued, and there are no
apparent criteria for determining which of the four values it should
bear. For these reasons, it seems that whilst the connection between

de Morgan implication and the four subsets of the set of the two classical truth values is mathematically neat, it is of no metaphysical significance.

We have already observed that the expression $p \wedge \neg p \to q \vee \neg q$ is not a thesis of de Morgan implication. Let us now define *Kalman implication* to be the relation between formulae determined by adding the scheme $\alpha \wedge \neg \alpha \to \beta \vee \neg \beta$ to the axiom system of de Morgan implication. Let us also define *the three element model* by the following diagram.

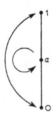

Exercise 51   Write down the tables for negation, conjunction and disjunction that correspond to this diagram.

Exercise 52   Verify that any expression of the form $\alpha \wedge \neg \alpha \to \beta \vee \neg \beta$ is valid in the three element model.

Exercise 53   Show that neither of the expressions $p \wedge \neg p \to q$ and $p \to q \vee \neg q$, where p and q are distinct propositional letters, is valid in the three element model.

Exercise 54   Show that the expression $(p \vee q) \wedge \neg p \to q$, where p and q are distinct propositional letters, is not valid in the three element model.

It turns out that validity in the three element model coincides with Kalman implication: an expression $\alpha \to \beta$ is a thesis of the axiom system for Kalman implication iff it is valid in the three element model. The soundness part of this theorem is easy to show: the central verification that is needed is already contained in exercise 52. The completeness part of the result can be obtained by making some small additions to the corresponding completeness proof for de Morgan implication.

Kalman implication can also be brought into relation with Aristotle's laws of thought. We can look on it as a logic that abandons either one, but not both, of the laws of contradiction and excluded third. We can treat the three values as corresponding to the three *proper* subsets of the set of the two classical values (retaining the law of contradiction and abandoning excluded third), or we may equally well regard the three values as corresponding to the three *non-empty* subsets of the set of the two classical values (retaining excluded third and jettisoning the law of contradiction). These two interpretations are indicated by the following diagrams:

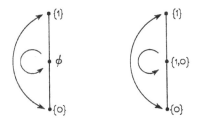

Once again, the correspondence is formally neat, but does not appear to be of philosophical significance.

Exercise 55   What kind of logic would correspond to abandoning Aristotle's law of identity?

De Morgan and Kalman implication have some other interesting formal properties. We mention two of them, without proof.

(1) De Morgan implication, Kalman implication and tautological implication diverge only in the vicinity of negation. As far as conjunction and disjunction considered alone are concerned, the three kinds of implication are indistinguishable from each other. More precisely, if α and β are formulae built up with $\wedge$ and $\vee$ but without any occurrences of $\neg$, then α de Morgan implies β iff α Kalman implies β, and also iff α tautologically implies β.

(2) Kalman implication and tautological implication between them exhaust the direct extensions of de Morgan implication. To put the matter precisely, if we add one or more axiom schemes to the axiom system for de Morgan implication, without changing the derivation rules, then the resulting implication relation must coincide

with one of the three already studied – de Morgan, Kalman, or tautological – or else be the 'total implication' under which each formula
implies every formula.

Up to this point, we have been treating implication as a *relation*
between formulae, and we have studied three such relations. However
logicians have also attempted to treat these kinds of implication as
operators, on an equal footing with $\neg, \wedge, \vee$. The basic idea is to deal
with formulae constructed by means of four operators $\neg, \wedge, \vee$, and
$\rightarrow$, or possibly others from which these four may be defined. Among
such formulae will be many in which one arrow occurs within the
scope of another. This happens, for example, in the formula

$$((p \rightarrow q) \wedge (q \rightarrow r)) \rightarrow (p \rightarrow r)$$

and to an even greater degree in the formula

$$(p \rightarrow q) \rightarrow ((q \rightarrow r) \rightarrow (p \rightarrow r)).$$

Attempts are made to define classes of 'acceptable' formulae, either
by means of axiom schemes and derivation rules, or by means of
models, in such a way as to obtain interesting generalizations of the
relations we have studied.

It is not difficult to transform the relation of tautological implication into an interesting operator. The transformation can be carried
out in several ways, but the most elegant is rather indirect. Instead of
beginning with the four operators $\neg, \wedge, \vee, \rightarrow$ we add a unary operator
$\Box$, where $\Box\alpha$ is read as 'it is necessary that α', to the classical operators
$\neg, \wedge, \vee$. We then set out various axiom schemes and derivation rules
to govern $\Box$, and introduce the arrow by definition, treating $\alpha \rightarrow \beta$ as
an abbreviation for $\Box(\alpha \supset \beta)$, and thus ultimately as an abbreviation
for $\Box(\neg(\alpha \wedge \neg\beta))$. Systems formed in this way are called *modal logics*.
There are a great many interesting modal logics, corresponding to the
axiom schemes and rules chosen for $\Box$, but two of the simplest among
them are the system M of Feys and von Wright, and the system S5 of
Lewis. For the system M we take as axioms all tautologies made up
using $\neg, \wedge, \vee$ and all their substitution instances, and also all formulae
of the form $\Box(\alpha \supset \beta) \supset (\Box\alpha \supset \Box\beta)$, and all formulae of the form
$\Box\alpha \supset \alpha$. We take as derivation rules for M the rule of material detach-
ment (from α and $\alpha \supset \beta$ to β) and the rule of necessitation (from α to
$\Box\alpha$). The system S5 is defined by adding the axiom schemes $\Box\alpha \supset \Box\Box\alpha$

and $\neg\Box\alpha \supset \Box\neg\Box\alpha$ to the above base. These systems, and others related to them, have received intensive study in recent years.

Systems have also been constructed that transform the relation of de Morgan implication into an operator. Here it is usual to proceed directly, taking the arrow as primitive along with $\neg$, $\wedge$, $\vee$. Several axiom systems have been constructed, and some of their properties have been studied. However these systems lack the simplicity and harmony of the modal logics. The list of axiom schemes is always long, and the rationale for the choice and exclusion of axiom schemes is tenuous. Moreover to date, no satisfying nontrivial characterizations of such systems have been given in terms of models. In the author's opinion, in the realm of nonclassical logic where nothing has usefulness to commend it, the only justification for existence is beauty, and systems transforming de Morgan implication into an operator lack it sadly. However many logicians would consider such views extreme.

3 Some aspects of quantificational logic

Outline

We begin by defining the relation of truistic implication, and setting out an axiom system that characterizes it. In doing so, we also explain the concepts of free and bound occurrences of individual variables, and the concept of a regular substitution. We then describe some of the deeper properties of quantificational logic, in particular the strong completeness theorem, compactness, the Löwenheim-Skolem theorem, and undecidability. The Löwenheim-Skolem theorem has some surprising and philosophically intriguing consequences for set theory, which are discussed in some detail. Finally, we explore some of the limits in the expressive capacities of quantificational logic, discussing in particular the relation of identity, functions, quantification over predicates, branching quantifiers, qualifiers, and proper modifiers.

1 An axiomatization of the relation of truistic implication

We recall that formulae of quantificational logic can be built from individual variables and predicate letters by means of the propositional operators $\neg$, $\wedge$, $\vee$, parentheses, and the universal quantifier $\forall$. The existential quantifier can be introduced by treating expressions of the form $\exists x(\alpha)$ as abbreviations for $\neg \forall x(\neg \alpha)$. It is possible to add further categories of symbols, such as function letters and a special binary predicate for the relation of identity, but for the present we shall omit them. Later we shall see that although these additional categories of symbol are in practice very convenient, they are also eliminable. Our first task is to define a relation that will play the same central role for quantificational logic as did the relation of tautological implication in truth-functional logic. The definition is rather complex in its details, and is most conveniently carried out using concepts of set theory. We begin by explaining the preliminary idea of an interpretation.

An *interpretation* of formulae of quantificational logic is made up of three parts. First: a set D, that is called the *domain of discourse* for the interpretation, and which contains at least one element. Second: a rule F that assigns values to predicate letters in the following ways—to each one-place predicate letter, F assigns a *subset* of the domain D; to each two-place predicate letter, F assigns a *binary relation* over the domain (in other words, a set of ordered pairs of elements of the domain); and in general to each n-place predicate letter, F assigns an n-ary relation over the domain (in other words, a set of ordered n-tuples of elements of the domain). Third: a rule f that assigns to each individual variable an element of the domain.

Now consider any interpretation, with a domain D, a rule F for assigning values to predicate letters, and a rule f for assigning values to individual variables. We define the truth-value of each formula upon that interpretation in the following inductive way.

(1) Let Px be an elementary formula made up of a one-place predicate letter P and an individual variable x. Then Px is said to be true under the interpretation if the element of the domain that the rule f assigns as the value of the variable x is an element of the set that the rule F assigns as the value of the predicate letter P. In other words, Px is true under the interpretation if $f(x)$ is an element of the set $F(P)$. More generally, let $Px_1 \ldots x_n$ be any elementary formula, made up of an n-place predicate letter, where n is any positive integer, and n individual variables $x_1, \ldots, x_n$ not necessarily distinct from each other. Then $Px_1 \ldots x_n$ is said to be true under the interpretation if the elements of the domain that the rule f assigns as values of the variables $x_1, \ldots, x_n$ stand to each other in the relation that the rule F assigns as the value of the predicate letter P. In other words, $Px_1 \ldots x_n$ is true under the interpretation if the ordered n-tuple $\langle f(x_1), \ldots, f(x_n) \rangle$ of values of the variables $x_1, \ldots, x_n$ is an element of the set $F(P)$ that is the value of the predicate letter P. If this condition is not satisfied, then $Px_1 \ldots x_n$ is said to be false under the interpretation.

(2.1) A formula of the form $\neg \alpha$ is true under the interpretation if α is false under the interpretation. Otherwise $\neg \alpha$ is false under the interpretation.

(2.2) A formula of the form $\alpha \wedge \beta$ is true under the interpretation if each of α and β is true under the interpretation. Otherwise $\alpha \wedge \beta$ is false under the interpretation.

(2.3) A formula of the form $\alpha \vee \beta$ is true under the interpretation if at least one of α and β is true under the interpretation. Otherwise $\alpha \vee \beta$ is false under the interpretation.

(3) A formula of the form $\forall x(\alpha)$ is true under the interpretation if α is true under every x-variant of the interpretation. Otherwise $\forall x(\alpha)$ is false under the interpretation.

Clearly conditions (2.1) through (2.3) are repeated from truth-functional logic. The new and interesting conditions are (1) and (3).

As an example to illustrate condition (1), suppose that our domain of discourse is the set of all positive integers, and that the one-place predicate letter P is assigned, by the rule F, the set of all odd integers as its value. Suppose moreover that the rule f assigns the value 2 to the individual variable x, and the value 3 to the individual variable y. Then the formula Py will come out as true under this interpretation, since 3 is an element of the set of all odd integers, in other words, since 3 is odd. On the other hand, the formula Px will come out as false under this interpretation, since 2 is not odd. Now suppose that the two-place predicate letter R is assigned as its value, by the rule F, the relation of being less than, in other words, the set of all ordered pairs $\langle m, n \rangle$ such that m is less than n. Then the formula Rxy will come out as true, since 2 is less than 3, in other words, since the pair $\langle f(x), f(y) \rangle$ is an element of $F(R)$. On the other hand the formula Ryx will come out as false, since 3 is not less than 2. Similarly, each of the formulae Rxx and Ryy will come out as false under this interpretation.

The terminology of condition (3) needs some explanation. Let x be any individual variable. Then one interpretation is said to be an *x-variant* of another iff the two agree in all respects except possibly the value that they assign to x. More explicitly, one interpretation is an x-variant of another iff the two interpretations share the same domain of discourse and the same rule for assigning values to predicate letters, and do not differ in the values that they give to any individual variables other than the variable x.

To illustrate this concept, suppose once more that our domain of discourse is the set of all positive integers, but suppose, this time, that the rule F assigns the set of all prime integers as the value of the one-place predicate letter P, and that the rule f assigns the value 7 to the individual variable x and the value 13 to the individual variable y. To help with cross reference in describing the example, we shall call this the first interpretation. Now we can form an x-variant of the first interpretation by giving the variable x the value 8, say, whilst holding unchanged the domain of discourse as well as the values of P, y, and all other predicates and individual variables. Moreover, although the formula Px comes out as true under the first interpretation, since the value of x is there 7, and 7 is prime, nevertheless the same formula Px comes out as false under the x-variant interpretation that we have described, since 8 is not prime. Thus by condition (3), the formula $\forall x(Px)$ comes out as false under the first interpretation (and indeed also under the second) since Px does not come out as true under every x-variant of that interpretation.

We can also form y-variants of our first interpretation, by choosing fresh values for the variable y whilst holding fixed the domain of discourse as well as the values of P, x and all other predicates and individual variables. For example, we might choose 10 as a value of y, whilst keeping all other values unchanged. Then the formula Py will come out as false under this y-variant of the first interpretation, and so by condition (3), the formula $\forall y(Py)$ comes out as false under the first interpretation itself.

As a curiosity, note that although the formulae $\forall x(Px)$ and $\forall y(Py)$ come out as false under the first interpretation, nevertheless the rather odd formulae $\forall x(Py)$ and $\forall y(Px)$, as well as, say, $\forall z(Px)$ and $\forall z(Py)$, all come out as true under the first interpretation. The reason for this is that in each case the variable that is bound by the quantifier is distinct from the variable that occurs in the body of the formula. Thus, to take the case of $\forall x(Py)$, whilst the x-variants of the first interpretation may shift the value of x at will, they all leave the value of the variable y untouched at 13. Hence the formula Py remains true under every x-variant of the first interpretation (though not, of course, under every y-variant of the first interpretation) so that by condition (3) the formula $\forall x(Py)$ comes out as true under the first interpretation.

Exercise 56 Let x be any individual variable. Verify that every interpretation is an *x*-variant of itself. Verify that if one interpretation is an *x*-variant of another, then that other is an *x*-variant of the first. Also verify that if one interpretation is an *x*-variant of a second, and the second is an *x*-variant of a third, then the first is an *x*-variant of the third.

Exercise 57   Using condition (3), state in a more 'positive' form the condition under which a formula of the form $\forall x(\alpha)$ is false under an interpretation.

Exercise 58   Given that $\exists x(\alpha)$ is an abbreviation for $\neg\forall x(\neg\alpha)$, state as simply as you can the condition for the truth of a formula of the form $\exists x(\alpha)$.

Exercise 59   Describe an interpretation under which the formula $\forall x(Px \lor Qx)$ is true, but under which the formula $\forall x(Px) \lor \forall x(Qx)$ is false. Set out in some detail the steps by which you calculate the truth values of these two formulae under the interpretation that you choose.

Exercise 60   Do the same for the pair of formulae $\exists x(Px) \land \exists x(Qx)$ and $\exists x(Px \land Qx)$, making use of the result of exercise 58.

Exercise 61   Do the same for the pair of formulae $\forall x(\exists y(Rxy))$ and $\exists y(\forall x(Rxy))$.

We now define the fundamental relation of truistic implication. If α and β are formulae, we say that α truistically implies β iff there is no interpretation under which α is true and β is false.

Exercise 62   Show that $\forall x(\alpha)$ truistically implies α, for any formula α. To do this, it is useful to follow a *'reductio ad absurdum'* procedure: suppose that $\forall x(\alpha)$ does not truistically imply α, and seek a contradiction.

Exercise 63   Show that $\forall x(\alpha \land \beta)$ truistically implies $\forall x(\alpha) \land \forall x(\beta)$, for any formulae α and β. Likewise for the converse. Follow the same procedure as in the preceding exercise.

Exercise 64   We know from exercise 59 that $\forall x(Px \lor Qx)$ does not truistically imply $\forall x(Px) \lor \forall x(Qx)$. Show that the latter formula does, however, truistically imply the former. Do the same for the pairs of formulae mentioned in exercises 60 and 61.

Exercise 65 Determine whether or not $\forall x(Px \supset Qx)$ truistically implies $\forall x(Px) \supset \forall x(Qx)$. Likewise for the converse.

Exercises 66 Determine whether or not $\forall x(\exists y(Rxy))$ truistically implies $\exists y(Ryy)$. Likewise for the converse.

We wish to axiomatize the relation of truistic implication. It turns out that we need only one more axiom scheme, and one more derivation rule, over and above those already occurring in our axiomatization of truth-functional logic. However, to state the scheme and derivation rule we need some terminology concerning freedom, bondage and substitution.

Let α be any formula and let x be any individual variable. We say that an occurrence of x is *bound* in α iff it is attached to a quantifier in α, or else occurs in the scope of a quantifier to which is attached another occurrence of x. We say that an occurrence of x is *free* in α iff it is not bound in α. The rough idea here is that an occurrence of an individual variable in a formula is called bound if it is worked upon by some quantifier of the formula. For example, the occurrences of individual variables that are bound in the following formula are indicated by lines.

$$\forall x\{(Px \wedge Py) \supset \exists y\,(Qy \wedge Rxy)\} \wedge Qx$$

From this example it is clear that a single variable may have one of its occurrences bound in a formula, and another of its occurrences free. In the above formula, the first three occurrences of x are bound whilst the fourth is free. The first occurrence of y is free whilst its other occurrences are bound.

Exercise 67 Indicate the occurrences of variables that are bound in the following formula. Be careful with parentheses.

$$\exists x\,\{\exists y(Rxz \wedge Rzy) \wedge \exists z(Rxy \wedge Ryz)\}$$

If α is a formula and x and y are individual variables, then we write $\alpha[x|y]$ to indicate the formula obtained by substituting the variable y for all the occurrences of x that are free in the formula α.

For example, let x, y, z be distinct individual variables, and let α be the formula $\exists z (Rxz)$. Then

$$\alpha[x\,|\,y] = \exists z (Rxz)[x\,|\,y] = \exists z (Ryz)$$
$$\alpha[x\,|\,z] = \exists z (Rxz)[x\,|\,z] = \exists z (Rzz).$$

Moreover we have

$$\alpha[x\,|\,x] = \exists z (Rxz)[x\,|\,x] = \exists z (Rxz)$$
$$\alpha[z\,|\,x] = \exists z (Rxz)[z\,|\,x] = \exists z (Rxz)$$
$$\alpha[y\,|\,x] = \exists z (Rxz)[y\,|\,x] = \exists z (Rxz).$$

The first two of these examples are quite straightforward. The formula $\exists z (Rxz)$ has just one free occurrence of x, and we simply put another variable in its place. However, the remaining examples need some explanation. In the third example, we replace the one free occurrence of x in $\exists z (Rxz)$ by x itself. Naturally this leaves us with the same formula as that with which we began. A substitution of this kind, where we replace the free occurrences of a variable by the very same variable, is called an *identity substitution.* In the fourth example, we are substituting x for all the occurrences of z that are free in α, but *no* occurrences of z are free in α, and so the result of the substitution just leaves us with α. Substitutions of this kind, where the variable being replaced has no free occurrences in the formula in question, are called *vacuous* or *empty* substitutions. Note that both identity and vacuous substitutions always leave us with the same formula as we began with. In the fifth example, the substitution is also vacuous.

Exercise 68   Give an example of an identity substitution that is also vacuous.

Let α be a formula and let x and y be individual variables. The substitution $\alpha[x\,|\,y]$ is said to be *regular* iff every occurrence of y that it introduces is free in the formula that results from the substitution. Thus in the preceding examples of substitutions, the second substitution is not regular. However, all the others are.

Exercise 69   Let x, y, z, be distinct individual variables and let α be the formula $y(Py \wedge Rxy)$. Which of the substitutions $\alpha[x\,|\,y]$ and $\alpha[x\,|\,z]$ are regular?

Exercise 70   Verify that identity and vacuous substitutions are always regular.

Exercise 71   Let α be the formula mentioned in exercise 67. Which of the following substitutions are regular: $\alpha[x\,|\,y]$, $\alpha[x\,|\,z]$, $\alpha[y\,|\,x]$, $\alpha[y\,|\,z]$, $\alpha[z\,|\,x]$, $\alpha[z\,|\,y]$?

We can now set out an axiom system for quantificational logic. We take all the axiom schemes and derivation rules of the system for truth-functional logic, and add the following ones.

Additional axiom scheme

$$\forall x(\alpha) \rightarrow \alpha[x\,|\,y]$$

Here α is any formula and x and y are any individual variables, not necessarily distinct from each other, *such that* the substitution $\alpha[x\,|\,y]$ is regular.

Additional derivation rule

From $\alpha \rightarrow \beta[x\,|\,y]$ to $\alpha \rightarrow \forall x(\beta)$

Here α and β are any formulae, and x and y are any individual variables, not necessarily distinct from each other, *such that* the substitution $\beta[x\,|\,y]$ is regular *and moreover* the variable y does not have any free occurrences in α.

Both the axiom scheme and the derivation rule are rather complex, and need some explanation.

The axiom scheme can be seen as a formal representation of the idea that 'every' implies 'any': if everything in a given domain of discourse satisfies a certain condition, then any particular thing that we care to choose from the domain will satisfy the condition. In formulating the scheme we have required that the substitution $\alpha[x\,|\,y]$ must be regular. This condition is irritating, but it (or some other device, such as distinguishing between two categories of variables) is necessary to avoid the variable y becoming entangled with other quantifiers already buried within α. An example will illustrate the point. We have already seen in exercise 66 that the formula $\forall x(\exists y(Rxy))$ does not truistically imply the formula $\exists y(Ryy)$. Thus if we want our axiom system to be sound with respect to the relation of truistic implication, then we should not allow as a thesis the expression

$$\forall x(\exists y(Rxy)) \to \exists y(Ryy).$$

But it is clear that if we take α to be the formula $\exists y(Rxy)$ then the expression displayed has the form

$$\forall x(\alpha) \to \alpha[x \mid y],$$

which coincides with that described in the axiom scheme, *except that* the occurrence of y that is introduced in place of the single free occurrence of x in α becomes bound by the existential quantifier in α, so that the substitution is not regular. We require regularity in the substitution in order to rule out cases like this.

When formulating the axiom scheme we allowed y to be the same variable as x. In such a case, we have a special and particularly simple form of the axiom scheme. For when y is the same variable as x, then the substitution is an identity substitution, the formula $\alpha[x \mid y]$ is just α, and the substitution is automatically regular. Thus the axiom scheme covers as a special case the following: the expression $\forall x(\alpha) \to \alpha$ is an axiom for any formula α and any individual variable x whatsoever.

The derivation rule for the quantifier reflects the traditional idea of proving a universal statement via an arbitrarily selected individual. This occurs frequently in school geometry, as in all other branches of mathematics. A geometer wishes to show that all triangles have a certain property, and reasons as follows. Let y be any triangle; for such-and-such reasons y has the property in question; so since y was an arbitrarily selected triangle, every triangle has the property in question. Of course, it is *vital* to the geometer's argument that the triangle y really is arbitrarily chosen and is not a special kind of triangle about which assumptions have been made in earlier hypotheses of the geometer's work. The derivation rule gives a precise syntactic content to the rather vague and psychological notion of an arbitrarily selected individual: the variable y must not have any free occurrences in the formula α.

Exercise 72 Formulate the special case of the derivation rule that arises when y is the same variable as x.

It is possible, and for some enjoyable, to spend a long time in becoming skilful in carrying out derivations in this axiom system. We shall concentrate on interests of a more theoretical kind, but a few derivations will help reveal the way in which the system works.

Exercise 73   Construct derivations in the axiom system for the scheme $\forall x(\alpha \wedge \beta) \rightarrow \forall x(\alpha) \wedge \forall x(\beta)$ and its converse. To do this, you will need only the special cases of the axiom scheme and derivation rule.

Exercise 74   Given that $\exists x$ is an abbreviation for $\neg \forall x \neg$, show that $\exists x(\alpha \vee \beta) \rightarrow \exists x(\alpha) \vee \exists x(\beta)$ and its converse are theses of the axiom system. Make use of the results of the preceding exercise.

Exercise 75   Show that if $\alpha \rightarrow \beta$ and $\beta \rightarrow \alpha$ are both theses of the axiom system, then so too are $\forall x(\alpha) \rightarrow \forall x(\beta)$ and $\forall x(\beta) \rightarrow \forall x(\alpha)$. Here too, only the special cases of the additional axiom scheme and derivation rule are needed.

Exercises 76   Show that $\forall x(\forall y(\alpha)) \rightarrow \forall y(\forall x(\alpha))$ is a thesis of the axiom system.

Exercise 77   Show that $\exists y(\forall x(\alpha)) \rightarrow \forall x(\exists y(\alpha))$ is a thesis of the axiom system. Begin by putting the expression in primitive notation.

Exercise 78   Let α be any formula and let x be an individual variable that does not occur free in α. Show that $\alpha \rightarrow \forall x(\alpha)$ is a thesis of the axiom system.

Exercise 79   Make use of the preceding exercise to show that $\forall x(\alpha) \rightarrow \forall x(\forall x(\alpha))$ is a thesis of the axiom system.

Exercise 80   Let α and β be formulae, and let x be an individual variable that does not occur free in α. Show that $\alpha \vee \forall x(\beta) \rightarrow \forall x(\alpha \vee \beta)$ and its converse are both theses of the axiom system. The converse part is difficult.

We now turn to questions of a more theoretical nature, beginning with soundness and completeness.

Theorem   Let α and β be any formulae of quantificational logic. Then α truistically implies β iff the expression $\alpha \rightarrow \beta$ is a thesis of the axiom system described. In other words, the axiom system is both sound and complete with respect to the relation of truistic implication.

We shall not go into the proof of this theorem, except to remark that it can be carried out by suitably elaborating the arguments used for soundness and completeness in truth-functional logic. The details of the elaboration involve some rather messy manipulations of individual variables.

The completeness theorem can also be proven in a rather stronger form, that speaks of sets of formulae rather than merely of individual formulae. Before we state the stronger version, we need to abstract our concepts to the appropriate level. Let X be any set of formulae, and let β be a formula. We say that the set X *truistically implies* β iff there is no interpretation that makes all the formulae in X true whilst it makes β false. We say that X *axiomatically yields* β iff there are formulae $\varphi_1, \ldots, \varphi_n$ in X such that the expression $(\varphi_1 \wedge \ldots \wedge \varphi_n) \to \beta$ is a thesis of the axiom system. Then *the strong form of the completeness theorem* can be put as follows.

Theorem   Let X be any set of formulae and let β be any formula. If X does not axiomatically yield β, then there is an interpretation whose domain is the set of all positive integers, under which all the formulae in X are true whilst β is false.

Exercise 81   Show that strong completeness immediately implies ordinary completeness.

The strong completeness theorem has two important corollaries, known as the *compactness theorem* and the *Löwenheim-Skolem theorem*.

First corollary   Let X be any set of formulae and let β be any formula. If X truistically implies β, then there is some finite subset X' of X such that X' truistically implies β.

This result is often summed up by saying that the relation of truistic implication is compact. The term 'compact' is used because of some analogies with topology. The result is of philosophical interest because it shows that, as far as quantificational logic is concerned, no valid argument requires an infinite collection of premises.

For if an argument is valid in quantificational logic, then the set of all its premises truistically implies its conclusion; and so even if the set of all the premises is infinite, nevertheless by the compactness theorem there will be *some finite subset* of these premises that truistically implies the conclusion.

Exercise 82   Show that compactness follows immediately from strong completeness taken together with soundness.

Second corollary   Let X be any set of formulae and let β be any formula. If X does not truistically imply β then there is an interpretation whose domain is the set of all positive integers, under which all the formulae in X are true whilst β is false.

Exercise 83   Indicate the sole point at which the formulation of this corollary differs from that of the strong completeness theorem. Show that the corollary follows immediately from strong completeness taken together with soundness.

The second corollary, known as the Löwenheim-Skolem theorem, has some philosophically intriguing consequences for set theory, as we shall explain in the next section.

2   *Some implications of the Löwenheim-Skolem theorem*

One of the fundamental classifications in set theory is that of finite and infinite. There are various equivalent ways of making the distinction, but one of the most useful is the following. A set is said to be *infinite* iff there is some one-one correspondence between it and one of its proper subsets, and a set is called *finite* iff it is not infinite. Thus for example, the set of all positive integers is infinite, since there is a one-one correspondence between it and, say, the set of all even positive integers. This correspondence can be described by the rule that associates each positive integer n with its double, $2n$, and can be pictured by means of a diagram.

$$1, \quad 2, \quad 3, \ldots, n, \ldots$$

$$2, \quad 4, \quad 6, \ldots, 2n, \ldots$$

The infinite sets admit of further classification. A set is said to be *denumerable* iff there is a one-one correspondence between it and the set of all positive integers. Thus, as the simplest example, the set of all positive integers is itself denumerable, for every set can be put into a one-one correspondence with itself. Again, the set of all even positive integers is denumerable, for we have just seen that there is a one-one correspondence between that set and the set of all positive integers. Again, the set of all integers, positive, zero and negative, is denumerable : the following diagram gives a picture of a one-one correspondence between this set and the set of all positive integers.

$$1, \quad 2, \quad 3, \quad 4, \quad 5, \quad 6, \ldots$$

$$0, \quad 1, \quad -1, \quad 2, -2, \quad 3, \ldots$$

Exercise 84   Describe in arithmetical terms the one-one correspondence pictured in the above diagram.

It can also be shown that the set of all rational numbers is denumerable. This result can be rather surprising because the natural ordering of the rational numbers, unlike that of the integers, is *dense* : between any two distinct rational numbers lies a third that is distinct from each of them.

Thus we have many examples of denumerable sets. What would be an example of an infinite set that is not denumerable? Write I for the set of all positive integers, and write s(I) for the set of all subsets of I. It is not difficult to verify that s(I), like I, is infinite. But it can also be shown that s(I) is not denumerable. In other words, there cannot be any one-one correspondence between s(I) and I itself. Indeed, we can show more generally that for *every* set X there is no one-one correspondence between X and the set $s(X)$ of all its subsets. The proof is due to Georg Cantor, towards the end of the nineteenth century, and forms a cornerstone of the theory of sets. It is a product of genius, but is of beautiful simplicity, as follows.

Suppose for *reductio ad absurdum* that there is a one-one correspondence between a set X and the set $s(X)$ of all its subsets. Call this correspondence g. Then with each element x of X, g associates an element $g(x)$ of $s(X)$. Now let Y be the set of all those elements x of X that are not elements of their images $g(x)$. Clearly, Y is a subset of X, and so Y is an element of $s(X)$. Hence, since g is a one-one correspondence between X and $s(X)$, there must be an element y of X such that $Y = g(y)$. Now we ask : is y an element of Y, or not? Suppose first that y is an element of Y. Then by the definition of Y, y is not an element of $g(y)$, which is Y, and so this supposition leads us to a contradiction. Suppose alternatively that y is *not* an element of Y. Then by the definition of Y, y is an element of $g(y)$, which is Y, and so this supposition also leads us to a contradiction. Since a contradiction arises in each of the two cases, we can say that the original supposition, that there is a one-one correspondence between X and $s(X)$, is false.

To sum up, in set theory we can prove that there are sets, such as $s(I)$, that are infinite but not denumerable. Now set theory itself can be set out in an axiomatic way. Moreover, we can choose a two-place predicate letter R, reading Rxy as 'x is an element of y', and if we wish, a one-place predicate letter S, reading Sx as 'x is a set', and set out the axioms of set theory as formulae of quantificational logic. The details need not concern us here, but the important point is that the collection of all the axioms of set theory can be envisaged as a certain collection of formulae in the notation of quantificational logic. Call this set of formulae X.

Now suppose that the collection X of formulae is consistent, that is, that X does not truistically imply any formula of the form $\beta \wedge \neg \beta$. Then we are faced with the following contrast. On the one hand, X truistically implies a formula α which, on its intended reading, says that there is an infinite set that is not denumerable. On the other hand, the Löwenheim-Skolem theorem tells us that there is an interpretation whose domain is the set of all positive integers, and so is denumerable, in which all the formulae in X, and so also α, are true. Thus the formula α asserts the existence of an infinite nondenumerable set, and yet is true under an interpretation with a denumerable domain. This situation is known as *Skolem's paradox*.

The first point to observe about Skolem's paradox is that it is not a contradiction. It does not reveal an inconsistency in logic or in axiomatic set theory. Nevertheless it is a strange and rather provoking phenomenon, that seems to hint at some kind of inadequacy or limitation in axiomatic set theory.

For example, Skolem himself has suggested that the paradox leads to a 'relativization' of the basic notions of set theory. Notions such as that of nondenumerability have no absolute meaning, but must always be understood as relative to a given system of axioms for set theory. A set may be shown to be nondenumerable within an axiom system, and yet be shown to be denumerable when viewed from outside the axiom system. Some writers have suggested a slightly different, but equally radical conclusion. From an absolute point of view, it is said, *all* infinite sets are denumerable, but from the limited point of view of any one axiom system for set theory, many among them can be proven not to be denumerable.

Other authors have approached the paradox from a quite different direction. For example, it has been suggested that the denumerable interpretations of set theory whose existence is guaranteed by the Löwenheim-Skolem theorem must be very complex and strange, and cannot bear any natural relation to the intended or desired inter-pretation. Thus if the domain of the denumerable model is the set of all positive integers, then the image of the elementhood symbol in the interpretation will be a complex and unwieldy relation between integers. For this reason, the denumerable interpretations of set theory may be regarded as quite pathological, with no real connection with the intuitive concept of a set.

Unfortunately neither of these points of view is entirely satis-factory. The second of the two is, in fact, inadequate for a purely technical reason. For although that point of view may be plausible as a way of looking at the *present* formulation of the Löwenheim-Skolem theorem, there are also other and *more powerful* versions of the theorem to which it does not apply. For example, it can be shown that whatever interpretation of set theory we might select to begin with, there is always a denumerable subinterpretation of it, in which exactly the same formulae are true as were true in the original interpretation. Thus, even if we leave aside the technical details involved in the definition of one interpretation being a subinterpretation of another, we can say that there are always denumerable models of set theory that bear a quite close relationship to whatever models of set theory we might initially regard as most natural.

Skolem's reflections on the paradox also face difficulties, but of a rather different kind. The suggestion that some or all of the concepts of set theory have only 'relative' meanings and no 'absolute' one excites the imagination, but is difficult to make clear or precise.

Indeed, at moments we seem to be grasping at something of great ontological importance, and yet when we close our hand we find that it escapes us.

However, even if Skolem's paradox does not lead to any clearly formulable ontological conclusions, it does suggest an interesting distinction between two kinds of formulae in set theory, which we now set out. As before, let X be a suitable set of axioms for set theory, in which the predicate letter R, say, is used to symbolize the relation of elementhood. In order to express the definition succinctly, let us follow the convention that if J is an interpretation with domain D, a rule F for predicate letters, and a rule f for individual variables, then we write $J = (D, F, f)$. Further, if $J = (D, F, f)$ is an interpretation and x is an individual variable then we write $f^*(x)$ to indicate the set of all elements of D that stand in the relation $F(R)$ to $f(x)$. Now let β be any formula in which just one individual variable x occurs free. We shall say that β is *stable* modulo X iff the following condition is satisfied: for every interpretation $J = (D, F, f)$, if β and all the formulae in X are true under J, then the set $f^*(x)$ has the property symbolized by the formula β. Otherwise β is said to be *unstable* modulo X.

Despite its complexity, this concept appears to arise quite naturally from the Skolem paradox. In effect, the paradox tells us that the formula of set theory that symbolizes the statement-form 'the set x is infinite but not denumerable', is unstable modulo the ordinary axiomatizations of set theory. On the other hand, the formula that expresses the statement-form 'the set x is infinite' is stable modulo the usual axiom systems for set theory. Thus one lesson of Skolem's paradox is to show the existence of unstable formulae in set theory. To be sure, the same lesson could have been learned from much more trivial examples. For instance, whereas the formula expressing the statement-form 'the set x has at least two elements' is stable modulo the usual axioms of set theory, it can be shown that the formula expressing 'the set x has at most two elements' is unstable modulo the same axiom systems. However nondenumerability gives us a particularly striking and unexpected case of instability. The concept of stability may perhaps merit further investigation.

Exercise 85 Generalize the concept of stability to apply to any formula in which one or more individual variables occur free.

Before leaving this section, we should mention another formal result of philosophical interest: the undecidability of quantificational logic.

A relation between formulae is said to be *decidable* iff there is some procedure, such that whenever we are presented with a pair α and β of formulae, we can apply the procedure in a completely mechanical way to find out, after a finite number of steps, whether or not α stands in the relation to β. Thus, for example, the relation of tautological implication is decidable. There is a procedure, namely the recipe for drawing up truth tables, such that whenever we take a pair α and β of formulae, we can apply the procedure to decide, after a finite number of steps, whether or not α tautologically implies β. In this case, the number of steps will vary, according to the degree of complexity of α and of β. If α and β have between them n propositional letters, the table will have 2^n rows. But in every instance, the application of the procedure is perfectly mechanical, and terminates with an answer. Again, the relation of de Morgan implication is decidable: there are four-valued truth tables that enable us to decide quite mechanically whether or not one formula de Morgan implies another.

It is natural to ask whether the relation of truistic implication is also decidable. The answer is no. The proof is of great depth, and we shall not attempt to summarize it here. Before the proof can even be begun, the concept of a decidable relation, which we have explained in only a vague and sketchy manner, needs to be given a precise mathematical definition. There are in fact several equivalent ways of doing this. One of these uses the idea of a 'recursive function' from the set of all positive integers into the same set, another introduces the idea of a perfect computer or 'Turing machine', and others proceed in yet different ways. The pursuit of these concepts leads us away from logic as initially and narrowly conceived, and into what is known as the theory of effective procedures.

The undecidability of quantificational logic is closely related to another result of philosophical interest: the essential incompleteness of arithmetic and set theory. It is possible to set out the ordinary arithmetic of the positive integers in an axiomatic way, and to set out these axioms as formulae in the notation of quantificational logic. If X is such a collection of axioms for arithmetic, then it can be shown that if X is consistent then X is *incomplete*, in the sense that there is a formula of formalized arithmetic, in which no individual variables occur free, such that neither it nor its negation is truistically implied by X. Moreover, X can be shown to be *essentially incomplete*, in the sense that every decidable but consistent set of axioms that includes X is also incomplete. The same results also hold true for set theory.

Here again, the proofs are deep and difficult. Their construction leads us into the general theory of the formalization of mathematical theories, or as it is sometimes known, metamathematics. Many of the results in this area, such as the incompleteness theorems mentioned above, stem from the work of Kurt Gödel.

3   *The expressive capacities of quantificational logic*

In this section we shall explore, by means of examples, the limits in the expressive power of quantificational logic. We shall begin by examining identity and functions, both of which can be regarded as 'positive' examples, and we shall then consider some 'negative' examples, such as quantification over predicates, parallel and branching quantifiers, and modifying adjectives.

There are two ways of approaching the expression of identity in the context of quantificational logic. One is to treat identity as a relation like any other, to be symbolized by a binary relation symbol. The truths concerning identity are not thought of as forming part of logic proper, but rather as providing a useful annex to logic. They are described by means of various axioms, and these axioms are regarded as forming the simplest and most universal part of mathematics.

The other approach is to treat identity as part of logic itself, by reducing it to indistinguishability. Suppose that in a certain context we are making use of only a finite number of primitive predicate letters. For example, if the subject is set theory, the primitive predicates might be just the one-place letter S and the two-place letter R, where Sx is read as 'x is a set' and Rxy is read as 'x is an element of y'. We can then define two individuals to be identical iff they are indistinguishable from each other by means of these two predicate letters, in other words, iff they are indistinguishable with respect to their sethood, with respect to the individuals to which they bear the relation of elementhood, and with respect to the individuals that bear the relation of elementhood to them. More briefly, in such a context we can define the expression $x = y$ to be an abbreviation for the conjunction

$$(Sx \equiv Sy) \wedge \forall z (Rxz \equiv Ryz) \wedge \forall z (Rzx \equiv Rzy).$$

If we follow this approach, then we have no need for any special postulates for identity. All the usual properties of identity can be derived from the axioms of quantificational logic, via the definition.

Thus identity may be reduced to indistinguishability in a context, and treated as part of logic, or it may be taken as a primitive concept and regarded as part of mathematics. The choice between the two is usually one of taste.

Exercise 86   Some systems of set theory dispense with the primitive predicate letter for sethood, and use only a binary predicate letter for elementhood. Express in symbols the definition of identity as indistinguishability in such a context.

Exercise 87   Express in symbols the definition of identity as indistinguishability in a context of just three predicate letters, one of which is one-place, another two-place, and the other three-place.

Exercise 88   What obstacle would be met in trying to treat identity as indistinguishability in the context of a theory that has infinitely many primitive predicate letters?

There are also two ways of expressing functions. We can add an entirely new category of symbols to represent them, or we can express them in terms of the categories of symbols that are already available. For example, in the context of arithmetic we could introduce a letter g to represent the successor function, and symbolize a statement such as 'there is a positive integer that is not the successor of any positive integer' as

$$\exists x (Ix \wedge \forall y (Iy \supset \neg (x = g(y)))).$$

On the other hand, we could also represent such a statement using only predicate letters. We could take a two-place predicate letter G, reading Gxy as 'y is a successor of x', describe the 'functionality' of this predicate by means of the formula

$$\forall x \, \forall y \, \forall z \, ((\mathrm{I}x \, \wedge \, \mathrm{I}y \, \wedge \, \mathrm{I}z \, \wedge \, Gxy \, \wedge \, Gxz) \supset y = z),$$

and express the arithmetical statement as

$$\exists x \, (\mathrm{I}x \, \wedge \, \forall y \, (\mathrm{I}y \supset \neg Gyx)).$$

This reduction transforms discourse with functions into discourse with predicates, and so reveals some of the expressive power of our minimum notation for quantificational logic.

> *Exercise 89*   Express the following statement of arithmetic, first with the help of a function letter for summation, and then without any function letters: 'the sum of x and y is always identical with the sum of y and x'.

We reach a limit, however, when we consider quantification over predicates. In so far as ordinary quantificational logic is concerned, the string $\exists x \, (Px)$ counts as a formula, but the string $\exists P \, (Px)$ does not. We might consider the possibility of extending quantificational logic by admitting quantification over predicates. We would then be able to study such interesting formulae as $\exists P(\forall x(\neg Px))$, $\forall P(\exists Q(\forall x \, (Qx \equiv \neg Px)))$, and the like. Unlike the extensions considered previously, this one cannot be expressed directly within the old notational resources. Such an extension provides us with what is known as *second-order logic*. Attractive as it can seem, however, the move turns out to be rather ill-advised, for the following reason. Second-order logic can in turn be extended to logics of higher and higher orders: we can add a new category of symbols, predicates of predicates, quantify over these, and so on in infinite sequence. Moreover the entire sequence of higher order logics can be looked upon as a mere fragment of set theory, and set theory itself can be expressed as a mathematical theory, with axioms of its own, but within the notation of ordinary quantificational logic. Thus all the advantages to be gained from second-order logic, and indeed more, can be gained by leaving the boundaries of logic alone, and studying set theory as a mathematical discipline.

Another possible extension of ordinary quantificational logic that has recently received attention is that of parallel or branching quantifiers. To understand the idea we need first to attend to the patterns of dependence involved in the ordinary use of quantifiers. We have already seen, in exercises 61 and 77, that there is an important difference between the formulae $\exists y (\forall x (Rxy))$ and $\forall x (\exists y (Rxy))$. The former truistically implies the latter, but not conversely. One rough way of putting the difference is by saying that in the latter formula, the choice of an appropriate y may depend upon the choice of x, whilst in the former, the choice of y is required to be independent of variations in x. Now consider a more complex formula,

$$\forall x (\exists y (\forall z (\exists w (Rxyzw)))).$$

Here the choice of y may depend upon x, and the choice of w may depend upon both x and z. It is natural to ask whether we can some-how make the choice of w depend upon, say, z alone, and not also upon x. In other words, we may ask whether there is any way of symbolizing a statement of the kind 'for every x there is a y, and for every z there is a w, depending on z alone, such that $Rxyzw$'. One way of doing this would be by allowing formulae to be composed by branching, as well as linear, arrangements of symbols. We could then write

$$\begin{array}{c} \forall x \exists y \\ \forall z \exists w \end{array} (Rxyzw).$$

Such an extension of quantificational logic has its attractions. It does not involve the addition of any new categories of symbols, but only the redeployment of old ones. But there is a reason for hesitating before the leap. The effect of branching quantifiers can be obtained by quantifying over functions. For example, the meaning of the branched statement written above can be represented as $\exists f \exists g \forall x \forall z (Rxf(x)zg(z))$. Now quantification over functions can be reduced to quantification over predicates, and thus to second-order logic. Moreover, second-order logic is best seen, as we have suggested, not as a part of logic at all, but as a fragment of the mathematical discipline of set theory. Thus while of interest in themselves, branching quantifiers belong more to mathematics than to logic.

Some interesting questions concerning the expressive power of quantificational logic also arise from a consideration of the idioms of natural language. For example, in everyday language we are continually applying adjectives to nouns, but we do so in two quite different ways, one of which is easy to symbolize in quantificational logic, and the other difficult.

Sometimes an adjective is thought of as representing a fixed property that does not vary according to the noun that follows it. More precisely, when the adjective is written before a noun or noun phrase, the resulting compound expression is taken to represent another property that is possessed by *just* those things that possess the two properties expressed by the adjective and the noun. Such adjectives will be called *qualifiers*.

When an adjective is used as a qualifier, then the set of individuals satisfying a compound expression made up of that adjective and a noun, is always a subset of the set of all individuals satisfying the noun considered alone. This property of qualifiers will be called *subsumption*. Moreover, when an adjective is used as a qualifier, its force is quite independent of the noun to which it is attached. To be precise, if an individual satisfies an expression made up of the adjective and a noun, and also satisfies another noun, then it satisfies the expression made up of the adjective and the second noun. This property of qualifiers will be called *transference*.

Purely qualifying adjectives occur quite frequently in mathematics, but quite rarely in everyday discourse. The adjective 'even' in arithmetic is a qualifier: a number is an even so-and-so just when it is both even and a so-and-so. It subsumes: for example, the set of all even products of seven is a subset of the set of all products of seven. It transfers: for example, if a number is an even product of seven, and is also a product of three, then it is an even product of three. In ordinary discourse, colour adjectives are paradigm qualifiers, or to be more accurate, paradigm approximations to qualifiers. Take the adjective 'blue'. It appears to be a qualifier: for example, an object is a blue pen if and only if it is blue, and also is a pen. We seem to have subsumption: for example, the set of all blue shirts is a subset of the set of shirts. There appears to be transfer: for example, if something is a blue tie and is also a museum exhibit then it is a blue museum exhibit. Yet even colour words may sometimes lack transference and so fail to be pure qualifiers. For example, we may taste something that is both a white wine and a liquid, but not a white liquid, but rather a light yellow one.

In cases like this, we are dealing with something more subtle than a qualifier. The adjective is not thought of as representing a fixed property, whose extension is unaffected by the nouns that follow it. A compound expression made up of adjective and noun is still regarded as indicating a property, but this property need not be possessed by exactly the individuals that possess properties expressed by the noun and the adjective separately. The adjective is thought of as *working on* the noun, rather than merely laid alongside it. When an adjective is used in such a way, it will be called a *modifier*. Qualifiers can be regarded as a special variety of modifiers, and those modifiers that are not qualifiers may be called *proper modifiers*. It should be added that a single adjective may play both roles; it may function as a qualifier in some of its uses, and as a proper modifier in others, as in the case of the colour word 'white'.

An adjective may fail to be a qualifier in a quite striking fashion, particularly when it lacks subsumption or transference. Subsumption is lost in the case of the set of all fake paintings of Picasso: it is not a subset of the set of all paintings of Picasso. Similarly with the set of all potential customers, former bachelors, decoy ducks, imitation diamonds, false teeth, assistant deans, apprentice mechanics, and alleged lawbreakers. Transference is lost with the adjective 'brilliant': a person may be a brilliant violinist and also a composer, without being a brilliant composer. As Aristotle remarked, a man may be a good painter or craftsman or fisherman and also be a citizen, without being a good citizen. A man may be a slow thinker and cyclist, yet not a slow cyclist.

Now logic has focused attention on qualifiers, to the exclusion of proper modifiers. This is natural enough, since in the modern period at least, logic has been very much concerned with describing and codifying the kinds of inference used in mathematical contexts, and in such contexts proper modifiers do not often occur. Indeed, when they do occur in mathematics, they are eliminable. But if we want to have a logic adapted to the idioms of ordinary language, we should perhaps look for some way of expressing modifiers.

One idea that has been discussed in recent years is to introduce a new category of symbols, *predicate operators*, into quantificational logic. A predicate operator will be a symbol which, when applied to a predicate, yields a predicate. Some work has been done in exploring the symbolization of modifiers as predicate operators, but the project has run into a technical difficulty which, so far as the author is aware, has not yet been overcome. If we are to build a semantics for such an extension of quantificational logic, then it seems natural to treat a one-place predicate operator as representing a function that takes every subset of the domain of discourse (and more generally, every n-adic relation over the domain) to another subset of the domain (or another n-adic relation). However this would have some undesirable effects. In particular, the formula $\forall x(Px \equiv Qx)$ would then logically imply the formula $\forall x(\pi(P)x \equiv \pi(Q)x)$, where π is any modifier, and this seems to clash with ordinary usage and understanding. Imagine, for example, a village in which there are just two fishermen and two barbers, and that these coincide. It may happen, however, that one of the two is a skilled fisherman and a clumsy barber, whilst the other is a skilled barber and a clumsy fisherman. In such a case, whilst it is true to say that the fishermen are just the barbers, it is not at all true to say that the skilled fishermen are just the skilled barbers.

Thus ordinary language seems to treat some proper modifiers in a nonextensional manner, and it is difficult to capture this treatment in a formal system of logic. The conjecture of the author is that the class of proper modifiers may be too broad and heterogeneous to admit of any single pattern of formalization. There may even be so many different kinds of proper modifier, each needing to be considered separately, that the logic of the subject may be as devious as, say, its grammar.

4 Remarks on the intuitionistic approach to logic

Outline

In this chapter we give a general account of the intuitionistic approach
to logic. Although, in the opinion of the author, the rationale for
this approach is quite misguided, it has been of some influence
among logicians in the present century.

Intuitionistic logic abandons some of the principles considered
correct in classical logic. We begin by indicating some of the most
striking of these, and setting out an axiomatic system that intuitionists
customarily accept as a satisfactory compendium of the principles of
logic that they retain. We then describe the general philosophical
ideas that provide the rationale for this pattern of rejection and re-
tention. Finally, we show how these philosophical ideas may be trans-
formed into rules for the understanding of each of the separate logical
operators.

1 Some patterns in intuitionistic logic

The first thing that a student hears about intuitionistic logic, and
usually the last thing that he forgets about it, is that it abandons some
of the principles of classical logic. For example, intuitionists accept
only part of the principle of double negation. Of the two implications

$$\neg\neg\alpha \to \alpha \qquad \alpha \to \neg\neg\alpha$$

they reject the left one, but accept the right one. Again, of the four
de Morgan principles

$$\neg(\alpha \wedge \beta) \to \neg\alpha \vee \neg\beta \qquad \neg\alpha \vee \neg\beta \to \neg(\alpha \wedge \beta)$$
$$\neg(\alpha \vee \beta) \to \neg\alpha \wedge \neg\beta \qquad \neg\alpha \wedge \neg\beta \to \neg(\alpha \vee \beta)$$

intuitionists abandon the top left one, but accept the remaining three.

A similar pattern arises in connection with the quantifiers. Of the four principles

$$\neg \forall x(\alpha) \rightarrow \exists x(\neg \alpha) \qquad \exists x(\neg \alpha) \rightarrow \neg \forall x(\alpha)$$
$$\neg \exists x(\alpha) \rightarrow \forall x(\neg \alpha) \qquad \forall x(\neg \alpha) \rightarrow \neg \exists x(\alpha)$$

the intuitionists drop the top left one and retain the remainder. Of the Lewis principles

$$\alpha \wedge \neg \alpha \rightarrow \beta \qquad \alpha \rightarrow \beta \vee \neg \beta$$

the intuitionists accept the former but not the latter. Whilst intuitionists accept the universal validity of the scheme $\neg(\alpha \wedge \neg \alpha)$ they deny that of the scheme $\beta \vee \neg \beta$. Whereas in classical logic some operators may be defined in terms of others, intuitionists regard all of $\neg, \wedge, \vee, \supset, \forall$, $\exists$ as primitive and as mutually independent. Most important of all, on the semantic level, intuitionists do not accept the idea, underlying classical logic, that every proposition is either true or false.

In one respect there is a resemblance between intuitionistic logic and de Morgan implication. Both question the law of excluded third, and both reject the principle $\alpha \rightarrow \beta \vee \neg \beta$. But in almost all other respects the two are in difference. For example, in rejecting one of the principles of double negation and one of the de Morgan principles, intuitionistic logic differs from de Morgan implication as much as it differs from tautological implication. Again, in accepting the Lewis principle $\alpha \wedge \neg \alpha \rightarrow \beta$ intuitionistic logic is more akin to classical logic than to de Morgan implication. Thus intuitionistic logic and de Morgan implication are two quite different departures from classical logic, and it would be a mistake to try to assimilate one of them to the other.

There is sufficient agreement among intuitionists as to which principles are correct, and which are incorrect, to enable the codification of intuitionistic logic as an axiom system. The following presentation of the system is neither the most common nor the most elegant, but it permits us to make a rapid comparison with the earlier axiomatization of classical logic.

Axiom schemes

$$\alpha \wedge \beta \rightarrow \alpha \qquad \alpha \rightarrow \alpha \vee \beta$$
$$\alpha \wedge \beta \rightarrow \beta \qquad \beta \rightarrow \alpha \vee \beta$$
$$\alpha \rightarrow \neg \neg \alpha$$

$$\alpha \to \alpha$$
$$\alpha \wedge (\beta \vee \gamma) \to (\alpha \wedge \beta) \vee (\alpha \wedge \gamma)$$
$$(\alpha \vee \beta) \wedge \neg \alpha \to \beta$$
$$\alpha \wedge (\alpha \supset \beta) \to \beta$$
$$\forall x(\alpha) \to \alpha[x|y] \qquad \alpha[x|y] \to \exists x(\alpha)$$

Derivation rules

From $\alpha \to \beta$	and $\beta \to \gamma$	to $\alpha \to \gamma$
From $\alpha \to \beta_1$	and $\alpha \to \beta_2$	to $\alpha \to \beta_1 \wedge \beta_2$
From $\alpha_1 \to \beta$	and $\alpha_2 \to \beta$	to $\alpha_1 \vee \alpha_2 \to \beta$
From $\alpha \wedge \beta \to \gamma \wedge \neg\gamma$ to	$\alpha \to \neg\beta$	
From $\alpha \wedge \beta \to \gamma$	to	$\alpha \to (\beta \supset \gamma)$
From $\alpha \to \beta[x|y]$	to	$\alpha \to \forall x(\beta)$
From $\beta[x|y] \to \alpha$	to	$\exists x(\beta) \to \alpha$

In the axiom schemes for $\forall$ and $\exists$ it is understood that the substitution $\alpha[x|y]$ is regular. In the derivation rules for $\forall$ and $\exists$ it is understood that the substitution $\beta[x|y]$ is regular and moreover that the individual variable y has no free occurrence in α.

If we compare this axiom system with the one for classical logic we see that, as would be expected from earlier remarks, one of the two axiom schemes for double negation is dropped whilst the other is retained. However, dropping $\neg\neg\alpha \to \alpha$ blocks the classical derivation of the principle of identity $\alpha \to \alpha$. Since the intuitionist has no objections to the principle of identity, this leads him to add $\alpha \to \alpha$ as a separate axiom scheme. The axiom schemes for conjunction, disjunction, distribution, and disjunctive syllogism are left unchanged. An axiom scheme of *modus ponens*, $\alpha \wedge (\alpha \supset \beta) \to \beta$ is added, since the intuitionist treats $\supset$ as a primitive operator. The axiom schemes for $\forall$ and $\exists$ are quite classical, but are given separately from each other, since the intuitionist does not take $\forall$ to be definable in terms of $\exists$ and $\neg$, or vice versa.

As far as the derivation rules are concerned, transitivity, conjunction in the consequent, and disjunction in the antecedent are retained without change. Contraposition is replaced by a form of *reductio ad absurdum*, authorizing the passage from $\alpha \wedge \beta \to \gamma \wedge \neg\gamma$ to $\alpha \to \neg\beta$. This is not because the intuitionist has any objections to contraposition in the form in which we have stated it. He accepts the validity

of moving from $\alpha \to \beta$ to $\neg\beta \to \neg\alpha$ although, we should add, he does
not accept the converse step from $\neg\beta \to \neg\alpha$ to $\alpha \to \beta$. The reason for
replacing contraposition by a form of *reductio ad absurdum* is purely
technical: in the context of weakenings in other parts of the system,
the rule of contraposition is not strong enough to obtain all that
the intuitionist wants from it, and so is replaced by a more powerful
rule.

Note at this point that we can distinguish between several different
forms of *reductio ad absurdum.* The differences between them are
quite trivial from a classical point of view, but of crucial importance
for the intuitionist. Both classicist and intuitionist accept the move,
already mentioned, from $\alpha \wedge \beta \to \gamma \wedge \neg\gamma$ to $\alpha \to \neg\beta$. The classical
logician also accepts *reductio ad absurdum* in the form that authorizes
passage from $\alpha \wedge \neg\beta \to \gamma \wedge \neg\gamma$ to $\alpha \to \beta$, and feels free to justify this
in terms of the first form, as follows: if we have $\alpha \wedge \neg\beta \to \gamma \wedge \neg\gamma$,
then by the first form of *reductio ad absurdum* we can proceed to
$\alpha \to \neg\neg\beta$, and since we know that $\neg\neg\beta \to \beta$ is a thesis, we can proceed
by transitivity to $\alpha \to \beta$. However this justification is not open to the
intuitionist, as he does not accept $\neg\neg\beta \to \beta$ as a thesis. The most that
the intuitionist can accept here is passage from $\alpha \wedge \neg\beta \to \gamma \wedge \neg\gamma$ to
$\alpha \to \neg\neg\beta$.

Reductio ad absurdum is closely related to a rule known as the
consequentia mirabilis, and a similar pattern arises here. One form of
the rule of *consequentia mirabilis* authorizes the step from $\alpha \wedge \beta \to \neg\beta$
to $\alpha \to \neg\beta$. This is accepted by classicists and intuitionists alike, both
of whom can justify it in terms of the first form of *reductio ad
absurdum* as follows: if we have $\alpha \wedge \beta \to \neg\beta$, then since we have the
thesis $\alpha \wedge \beta \to \beta$ we may proceed by conjunction in the consequent to
$\alpha \wedge \beta \to \beta \wedge \neg\beta$ and from this by the first form of *reductio ad absurdum*
to $\alpha \to \neg\beta$. However, *consequentia mirabilis* also has a second form,
authorizing passage from $\alpha \wedge \neg\beta \to \beta$ to $\alpha \to \beta$. This is acceptable
classically, but not intuitionistically.

Returning now to the rules that figure as derivation rules of the
axiom system, we can say that the rule of moving from $\alpha \wedge \beta \to \gamma$ to
$\alpha \to (\beta \supset \gamma)$, sometimes called exportation, is needed to handle the
operator $\supset$. The final two rules, for $\forall$ and $\exists$, are quite classical, as
were their corresponding axiom schemes, but both rules are needed
because $\forall$ is not defined from $\exists$ and $\neg$, or vice versa.

Exercise 90 Construct a derivation in the axiom system for each of the three de Morgan principles that were said to be accepted by intuitionists.

Exercise 91 Construct a derivation in the axiom system for each of the three quantifier negation principles that were said to be accepted by intuitionists.

Exercise 92 Show that the intuitionistic logic satisfies the rule of contraposition. In other words, show that whenever $\alpha \to \beta$ is a thesis, then so too is $\neg\beta \to \neg\alpha$.

Exercise 93 Construct a derivation in the axiom system for the one Lewis principle that was said to be accepted by intuitionists.

2 *Philosophical ideas underlying intuitionistic logic*

What is the rationale behind the curious pattern of intuitionistic logic? What leads the intuitionist to reject some of the principles of classical logic, but not others? The basis lies in the meanings that the intuitionist gives to the logical operators.

In classical logic we take the meaning of a logical operator to be determined by the conditions for the truth and falsehood of compound statements involving it. The simplest example is negation. A formula of the form $\neg\alpha$ is taken to be true under a given interpretation iff α is false under that interpretation, and moreover this stipulation is usually understood as providing the meaning of negation.

However intuitionists take the view that we cannot really ascribe truth and falsehood, as objective qualities, to the statements of mathematics. They concede that this may be possible with statements of everyday life that describe the physical world around us, but hold that when it comes to mathematics, the concepts of truth and falsehood do not have their usual application. Mathematical statements do not describe a reality existing independently of human activities. Instead, they describe the mental procedures and constructions of mathematicians themselves, or some idealization of them.

Now if we cannot ascribe truth and falsehood to mathematical statements, the idea continues, we should not determine the meaning of a logical operator occurring in a mathematical context by means of conditions for the truth and falsehood of compounds involving the operator. Since mathematical statements describe the mental constructions of real or idealized mathematicians, the meaning of each logical operator occurring in a mathematical context should be determined by the conditions under which a construction would count as an intuitively acceptable proof of compounds involving the operator. In brief: within a mathematical context, the meaning of a logical operator is determined not by the *truth conditions* but by the *provability conditions* of compounds involving it.

In the opinion of the author, these ideas are misguided. Even granting that mathematical statements considered in themselves have no status as true or false, and do not describe a special 'mathematical reality' independent of ourselves and the world about us, it does not follow that mathematical statements describe the mental constructions of the mathematicians, real or idealized, who envisage them. Nor does it follow that the meaning of a logical operator in a mathematical statement may not be understood in terms of truth conditions. These points become clearer if we fix our attention upon a particular branch of mathematics, and the simplest one to consider is geometry.

The axioms of geometry need not be understood, as Plato took them, as descriptions of a transcendental world of points without size, lines without width, absolutely perfect circles, and so on, existing independently of the human race and indeed beyond the confines of the physical universe. We have no need for such metaphysical hypotheses. Nevertheless, we are not obliged by this to conclude that the axioms of geometry derive their significance as descriptions of the workings of the mind of the geometer. A much more plausible view is available. The axioms of geometry set out certain general patterns and relationships. We do not find in the world around us any objects that answer *exactly* to these patterns. But when we interpret the primitive concepts of geometry in appropriate ways, we do find in the physical world objects and relationships that provide very good approximations to the axioms. We are then in a position to make use of the theorems of geometry, for as they follow logically from the axioms, they too will usually be good approximations.

Ideas such as these, in the view of the author, provide the most satisfactory view not only of geometry but also of other branches of mathematics. In the next chapter we shall consider their application to set theory. However these criticisms are leading away from the exposition of intuitionism. On the intuitionistic view, whether right or mistaken, the meaning of a logical operator in a mathematical context should be determined not by the truth conditions of compounds involving it, but instead by their provability conditions. We now look at the attempt to give content to this rather vague pronouncement.

3 The intuitionistic account of the logical operators

How can we unfold the general pronouncements of intuitionism so as to uncover the behaviour of each of the separate logical operators? One approach, that stems from the work of the intuitionists Brouwer and Heyting, and which has been developed by Kreisel and Troelstra, makes use of the notion of a 'construction'. The meaning of each of the logical operators $\wedge, \vee, \supset, \neg, \forall, \exists$ is given by a rule.

Rule for conjunction A construction is a proof of a statement of the form $\alpha \wedge \beta$ iff it is made up of two constructions, one of which is a proof of α and the other a proof of β.

Rule for disjunction A construction is a proof of a statement of the form $\alpha \vee \beta$ iff it is a proof of α or it is a proof of β.

Rule for the conditional A construction is a proof of a statement of the form $\alpha \supset \beta$ iff it is made up of two constructions, one of which provides an effective procedure for transforming any proof of α into a proof of β, and the other of which is a proof of that fact.

Rule for negation A construction is a proof of a statement of the form $\neg \alpha$ iff it is made up of two constructions, one of which provides an effective procedure for transforming any proof of α into a proof of some arbitrarily chosen absurd statement (say of $0 = 1$ if the subject matter is arithmetic), and the other of which is a proof of that fact.

Rule for the universal quantifier A construction is a proof of a statement of the form $\forall x(\alpha)$, in a given mathematical context, iff it is made up of two constructions, one of which provides an effective procedure for constructing a proof of $\alpha[x|a]$ for every individual constant *a* of the context, and the other of which is a proof of that fact.

Rule for the existential quantifier   A construction is a proof of a statement of the form $\exists x(\alpha)$, in a given mathematical context, iff it is a proof of $\alpha[x|a]$ for some individual constant *a* of the context.

Accompanying these rules for the separate logical operators is a general definition of intuitionistic validity. Let α be any formula, built using the predicate letters $P_1, \ldots, P_n$. Then α is said to be *intuitionistically valid* iff for every choice of a domain of discourse, and every choice of properties of elements of the domain as values for $P_1, \ldots, P_n$ there is a construction that is a proof of the statement that is thus correlated with α. In the case of formulae of propositional logic, built from propositional letters by means of the operators $\wedge$, $\vee$, $\supset$, $\neg$ and without quantifiers, the definition of intuitionistic validity can be simplified a little. Let α be any such formula, whose propositional letters are $p_1, \ldots, p_n$. Then α is said to be intuitionistically valid iff for every choice of statements as values of $p_1, \ldots, p_n$ there is a construction that is a proof of the statement thus correlated with α. There are several points here that deserve comment.

(1) The definition of intuitionistic validity is rather guarded in its use of the language of set theory. The definition does indeed speak of domains of discourse, but phrases such as 'subset of the domain' and 'set of ordered *k*-tuples of elements of the domain', which occupy a central place in the corresponding classical definitions, are supplanted by terms such as 'property', which the classical logician would regard as disagreeably vague. The reason for this difference is that intuitionists object not only to aspects of classical logic, but also to various branches of classical mathematics. In particular, intuitionists do not accept the concept of a set in the way in which it is classically formulated, and so are not prepared to make the same free use of set-theoretic terminology as are classical mathematicians, when defining or discussing concepts of logic.

(2) In the rules for the quantifiers, the notion of an individual constant plays a central role. In general, one can say that the concept of an individual constant is much more important for the understanding of intuitionistic logic than it is for the semantics of classical logic.

(3) Several of the rules – for the conditional, negation, and the

universal quantifier – make use of the concept of an effective procedure. By an effective procedure the intuitionist means a finite
series of instructions that can be followed through in a quite mechanical way to achieve a certain result. Of course the notion needs further
explanation to make it precise. It is closely related to the classical
concept of an effective procedure, mentioned in the preceding
chapter, but in general intuitionists deny that the classical concept
gives exactly the idea that they have in mind.

(4) Each of the rules that refers to an effective procedure has a
further complexity. As well as requiring the presence of an effective
procedure of a certain kind, it requires the presence of another construction that is a proof that the procedure *is* effective and *is* of the
kind desired.

The purpose of this last rather subtle complication becomes clearer
when we see its application to the formula $p \lor \neg p$ which, as we have
already remarked, intuitionists do not accept as correct. Suppose that
in the rule for intuitionistic negation we were to omit the requirement
of a second construction. Then if we were to use classical logic in
applying the rules and definitions, we would be led to classify $p \lor \neg p$
as intuitionistically valid, for we would be able to argue as follows:

Take any proposition p. Now either there is a construction that is
a proof of p or there is not. If there is such a construction, then by
the rule for disjunction, it is a proof of $p \lor \neg p$. If there is not any
construction that is a proof of p, then clearly the construction of
writing down nothing at all (the empty construction) is an effective
procedure for transforming any proof of p into a proof of say
$0 = 1$, and so by the rule for negation in its truncated form, the
construction of writing down nothing at all is a proof of $\neg p$, and
so in turn is a proof of $p \lor \neg p$.

When the rule for negation is stated in its full form, without
omission, then such an argument breaks down. For even if the procedure of writing down nothing at all constitutes an effective procedure for transforming any proof of p into a proof of $0 = 1$, we do
not have a *proof* that it does so unless we already have a proof that
there is no construction that is a proof of p, and in general this will be
lacking. To sum up: in the rule for negation, and each of the other
rules that makes use of the notion of an effective procedure, a second

clause is included, so that when we apply the rule with the help of classical reasoning, we obtain the same result as when we apply it with the help of intuitionistic reasoning.

The intuitionistic rules for the logical operators still have several areas of vagueness. As we have already remarked, they make use of the notions of a property, a construction, a proof, and an effective procedure. For this reason we can say that the intuitionistic account of the logical operators is still rather rough and incomplete. Despite this, however, the rules serve to give some insight into the status of individual formulae. For example, we have already seen that when properly understood, the rules provide a basis for rejecting the law of excluded third. The same is true of the examples in the following five exercises.

Exercise 94   The formula $(p \supset q) \lor (q \supset p)$ is a classical tautology. Explain, in the light of the intuitionistic rules for the logical operators, why intuitionists do not accept it as valid.

For the following exercises we need a further definition. An expression of the form $\alpha \to \beta$ is said to be intuitionistically acceptable iff the corresponding conditional formula $\alpha \supset \beta$ is intuitionistically valid.

Exercise 95   Explain, in the light of the intuitionistic rules for the logical operators, why intuitionists do not regard the expression $\neg(p \land q) \to \neg p \lor \neg q$ as acceptable.

Exercise 96   Do the same for the expression $\neg \forall x(Px) \to \exists x(\neg Px)$.

Exercise 97   Do the same for the expression $\neg\neg p \to p$.

Exercise 98   Likewise for the expression $\forall x(Pa \lor Qx) \to Pa \lor \forall x(Qx)$. Compare this with the result of exercise 80.

There is a criticism that has sometimes been made of the intuitionistic account of the logical operators, and which is much more radical than the charge of being rough and incomplete. The intuitionistic rules themselves *make use* of logical operators 'iff', 'and', 'or', 'every', and 'some'. In itself, this is not unusual. In classical logic, the rules for determining the truth value of a compound statement under a given interpretation make use of operators corresponding to the very ones that they are describing. But in the present situation, the use of

operators raises a delicate question. We can ask whether, in the intuitionistic rules, the operators that are *being used* are to be understood
classically, or intuitionistically. This question edges the intuitionist
towards a dilemma. If the operators that are used in the rules are
themselves to be understood intuitionistically, then the rules seem to
do little to explain the meanings of the intuitionistic connectives to a
person accustomed to classical logic. If on the other hand the operators that are used in the rules are to be taken classically, then the
intuitionist seems to be making use of the very logic that he professes
to abandon.

Intuitionism has an answer for this dilemma. In certain special
situations, the intuitionists suggest, when we are dealing only with
'intuitively decidable' propositions, classical logic may legitimately be
used. A statement is called intuitively decidable iff either we have a
proof of it, or else we have a proof that any proof of it would yield a
proof of any proposition whatsoever. Furthermore, the intuitionist
suggests, a statement of the form 'the construction π is a proof of the
proposition γ' is always decidable: given any construction π and any
statement γ, either we have a proof that π is a proof of γ, or else we
have a proof that any proof that π is a proof of γ would yield a proof
of any proposition whatsoever. Thus, taking certain particular cases
of γ, statements of the following forms are intuitively decidable: 'the
construction π is a proof of the proposition that the construction σ
provides an effective procedure for transforming any proof of α into
a proof of β', and 'the construction π is a proof of the proposition
that the construction σ provides an effective procedure for construct-
ing a proof of $\alpha[x|a]$ for every individual constant a'. For these
reasons, the intuitionist continues, the phrasing of the rules for the
logical operators falls within the limited area in which we may legitimately speak classically. In this way the dilemma appears to be by-
passed and the criticism answered.

Intuitionism is a difficult and often tricky subject. A fuller account
of intuitionism would have to consider not only logic but also mathematics. Whereas in logic the principles accepted by the intuitionists
form a subset of those accepted classically, in mathematics the situation is very much more complex. There are concepts and principles
of classical mathematics that the intuitionists reject, or accept only
in modified form. But the reverse also arises: intuitionistic

mathematics makes use of concepts, such as that of a free choice sequence, that have no place in classical mathematics, and even when the intuitionistic and classical mathematicians appear to be speaking about the same subject, there are occasions when the intuitionist will accept a principle that the classical mathematician rejects.

However, in this chapter we have restricted attention to intuitionistic logic. On the one hand, we have seen how some of the objections that arise naturally against it may be answered. In particular, the attempt to force the intuitionist into the dilemma of choosing between classical logic and incommunicability is inconclusive. On the other hand, we have also seen that there are objections, of a less radical kind, that are more difficult to answer. They can be summarized as follows:

(1) There is no need for a special intuitionistic logic, since the general philosophical ideas concerning the meaning of mathematical statements, which provide its rationale, appear to be misguided;

(2) The description of intuitionistic logic that is available to us today still has residual areas of vagueness, particularly in the notions of a property, construction, proof, and effective procedure, and so is rather rough and incomplete;

(3) The structure of intuitionistic logic is tiresomely complex. Good reasons would be necessary to force us to abandon its simpler and stronger classical counterpart.

5 From logic to set theory

Outline

We begin with a sketch of one of the more important axiomatizations of set theory, due to Zermelo. We then trace the web of interconnections between logic on the one hand and set theory on the other, and discuss the nature of set theory itself.

1 A sketch of Zermelo's axiomatization of set theory

In the first chapter of this book we introduced some of the basic concepts of set theory as tools to be used in studying systems of logic. The presentation was loose and unsystematic, as attention was on applications rather than on the theory itself. We shall now look at set theory from a more abstract perspective.

There are several different ways in which the theory of sets may be axiomatized. Some of the axiom systems are so distinct from each other as to be mutually incompatible. For example, one system, devised by W. V. Quine, implies the existence of a set of all sets, whilst most others contain axioms that imply the non-existence of such a set. We shall describe one of the more important and widely used axiomatizations of set theory. It was first constructed by E. Zermelo in the early years of the twentieth century, and has since then received various refinements and additions. This system can be seen as a point of departure for several others, such as those of von Neumann, Bernays, and Gödel, which, although they differ in various ways from Zermelo's axiomatization, can nevertheless be seen as developments and modifications of it.

Even within the context of Zermelo's approach, several

variations are possible, according to whether we include or omit various marginal axioms. There is also some room for choice in the treatment of the relation of identity, and in the choice of primitive predicates. It is customary to seek formal simplicity by using only one primitive predicate, the binary symbol $\in$ for elementhood. However for perspicuity we shall add a further one-place predicate, the symbol S for sethood. We shall treat identity as indistinguishability in the context of $\in$ and S, along the lines indicated in the third chapter. However these are minor matters, and the main ideas involved in Zermelo's axioms can be understood without too much attention to them.

The first axiom gives a criterion for the identity of sets. It says in effect that coextensive sets are identical.

Axiom of extensionality   If two sets have exactly the same elements, then they are identical. In symbols,

$$\forall x \forall y((Sx \land Sy \land \forall z(z \in x \equiv z \in y)) \supset x = y).$$

The next axiom is an existence statement. It asserts, quite unconditionally, the existence of an empty set.

Axiom of an empty set   There is a set that has no elements at all. In symbols,

$$\exists x(Sx \land \forall y \urcorner (y \in x)).$$

Exercise 99   Use the axiom of extensionality to show that there is at most one empty set. In this and the subsequent exercises, proceed informally, using the *verbal* formulation of the axioms concerned, rather than their symbolic expressions.

The next group of axioms can be described as conditional existence statements. In effect, they indicate ways of generating new things from old.

Axiom of pairs   For any objects x and y, not necessarily distinct, there is a set z whose elements are just x and y. In symbols,

$$\forall x \forall y \exists z (Sz \land \forall w(w \in z \equiv (w = x \lor w = y))).$$

Exercise 100   Use the axiom of pairs to show that for every object x there is a set whose sole element is x.

Axiom of union For every set x there is a set y whose elements are just those things that are elements of at least one of the elements of x. In symbols,

$$\forall x(Sx \supset \exists y(Sy \wedge \forall z(z \in y \equiv \exists w(w \in x \wedge z \in w)))).$$

Exercise 101 Use the axiom of pairs and the axiom of union to show that for any sets x and y there is a set z whose elements are just those things that are elements of at least one of x and y.

Axiom of powersets For every set x there is a set y whose elements are just the subsets of the first. In symbols, and using $z \subseteq x$ as an abbreviation for $\forall w(w \in z \supset w \in x)$, this axiom can be written

$$\forall x(Sx \supset \exists y(Sy \wedge \forall z(z \in y \equiv (Sz \wedge z \subseteq x)))).$$

Axiom scheme of separation Let α be any formula whose free variables are just $x_1, \ldots, x_n, z$. Then for all objects $x_1, \ldots, x_n$ and for every set x, there is a set y whose elements are just those objects z that are elements of x and satisfy the formula α.

In symbols,

$$\forall x_1 \ldots \forall x_n \forall x(Sx \supset \exists y(Sy \wedge \forall z(z \in y \equiv (z \in x \wedge \alpha)))).$$

To be exact, this is an axiom scheme rather than a single axiom, for it covers infinitely many distinct formulae, one for each choice of a formula α, and each of these formulae counts as a distinct axiom.

The scheme of separation is rather more complex in its formulation than any of the preceding axioms. However the basic idea behind it is quite simple. Forget for a moment the universally quantified variables $x_1, \ldots, x_n$; in most of the simpler applications of the scheme, n is taken to be zero. Then the scheme tells us, in effect, that every property marks off a subset of any given set. In other words, given any property expressible in the notation of our system of set theory, and given any set, there is a set whose elements are exactly the elements of the first set that possess the property.

Exercise 102 Use the axiom of union and the axiom scheme of separation to show that for any non-empty set x there is a set y whose elements are just those things that are elements of every one of the elements of x.

Exercise 103 Use the axiom of pairs and the result of the preceding exercise to show that for any sets x and y, there is a set z whose elements are just those things that are elements of both x and y.

In the early years of set theory, in the late nineteenth century, it was generally assumed that the idea of the scheme of separation could be set out in a simpler and more powerful form: every property determines a set. In other words, given any property expressible in the notation of set theory, there is a set whose elements are exactly those things that possess the property. Such a principle is known as the *scheme of comprehension.* If accepted, it would make several of the other axioms of set theory redundant; in particular, it would imply as special cases the axiom of pairs, the axiom of union, and the axiom of powersets. Unfortunately, however, this scheme leads to a contradiction, and so cannot be accepted. For it follows from the scheme of comprehension that there is a set x whose elements are just those things that possess the property of not being elements of themselves. Then in particular, x will be an element of x just if x is not an element of itself. So if x is an element of itself then it is not; and if it is not an element of itself then it is: contradiction. This contradiction was discovered by Bertrand Russell and is usually known as *Russell's antinomy.* The rather complex structure of the scheme of separation, in which the application of a property marks off only the *subsets* of given sets, can be seen as a means of avoiding the contradiction that is implicit in the simpler scheme of comprehension.

Exercise 104 Use the axiom scheme of separation to show that there is no set of all sets. To do this, suppose for *reductio ad absurdum* that there is such a set, and seek a contradiction.

Exercise 105 Use the results of exercises 104 and 101 to show that absolute complements never exist; in other words, that for no set x is there a set z whose elements are just those things that are not elements of x.

Exercise 106   Use the axiom scheme of separation to show that relative complements always exist; in other words, that for all sets x and y, there is a set z whose elements are just those elements of y that are not elements of x.

Exercise 107   Use the axiom scheme of separation to show that for every set x there is a subset of x that is not an element of x. This can be seen as a generalization of exercise 104, and can be approached in a similar manner.

Finally, we give another existence statement as an axiom. It asserts in effect the existence of an infinite set.

Axiom of infinity   There is a set x such that the empty set is an element of x and such that whenever an object y is an element of x, then so too is the set whose only element is y. To put this in symbols, first introduce $\phi \in x$ as an abbreviation for

$$\forall z((Sz \wedge \forall w \neg (w \in z)) \supset z \in x)$$

and use $\{y\} \in x$ as an abbreviation for

$$\forall z((Sz \wedge \forall w(w \in z \equiv w = y)) \supset z \in x).$$

Then we can write the axiom of infinity itself as

$$\exists x(Sx \wedge \phi \in x \wedge \forall y(y \in x \supset \{y\} \in x)).$$

To this basic list of axioms, others are sometimes added. The further axioms are rather complex in their exact formulation, and we shall merely sketch them. One among them is the *axiom scheme of replacement*, which is a generalization of the axiom scheme of separation. When Zermelo's axiom system is supplemented by this axiom scheme, it is usually known as *Zermelo-Fraenkel* set theory. At times, various generalizations of the axiom of infinity known as *strong axioms of infinity* are added. Another axiom that is often added is the *axiom of choice*. This can be put in words as follows. Let x be any set whose elements are all non-empty sets, and such that no two elements of x have any elements in common. Then there is a set z such that for each element y of x, one and only one element of y is an element of z.

Exercise 108   In formulating the axiom of choice we have required that the elements of x must all be non-empty. Explain why this condition is necessary.

Exercise 109   In formulating the axiom of choice we have required that no two distinct elements of x have any elements in common. Explain, with the help of a simple example, why this condition is necessary.

Still another principle that is sometimes considered as an axiom is the *principle of foundation.* This can roughly be described as a nonexistence principle. It says that there is no infinite sequence of sets $x_1, x_2, x_3, \ldots$ such that $x_{i+1} \in x_i$ for every $i \geqslant 1$. In other words, there are no infinite descending chains of the kind

$$\ldots \in x_{i+1} \in x_i \in \ldots \in x_2 \in x_1.$$

The axiom foundation is conceptually attractive. However it has little application within set theory; none of the theorems that mathematicians usually regard as central to set theory depend upon it.

Exercise 110   Use the axiom of foundation to show that no set is an element of itself.

Exercise 111   Use the axiom of foundation to show that there are no sets x and y such that x is an element of y and y is an element of x.

Exercise 112   Use the axiom of foundation to show that there is no finite sequence $x_1, \ldots, x_n$ of sets such that $x_1 \in x_2 \in \ldots \in x_n \in x_1$.

2   Interconnections between logic and set theory

The relationship between logic and set theory is a delicate one. Indeed, the distinction between the two has not always been clear. For example, in the early years of modern logic, in the late nineteenth century, Frege treated as part of logic material that nowadays would usually be regarded as a variant form of set theory. Following in Frege's footsteps, Whitehead and Russell announced in the early years of the twentieth century the reduction of all of classical mathematics to logic. Today, however, we speak only of the reduction of mathematics to set theory, and abandon the attempt, made by Whitehead and Russell, to reduce the concept of a set to that of a propositional function and to insert the latter within logic itself. As was remarked in an earlier chapter, such devices as quantification over predicates, higher order logics, and the simple and ramified type theories of Whitehead and Russell, tend now to be regarded as setting out, often in a rather indirect form, various fragments of set theory. In general, it has become more and more widely accepted that quite a sharp distinction can be drawn between logic on the one hand, and set theory, arithmetic, and the remainder of mathematics on the other, and that the most natural point to make the break is quite low on the scale, at the very perimeter of ordinary quantificational logic.

Granted that there is a distinction between logic and set theory, the interconnections between the two still need clarification. On the one hand set theory, like any other mathematical theory, makes use of logic. It uses the operators of logic, and in so far as it is formalized, employs the notations of logic. It makes use of logic in deducing its theorems from its axioms, and the kind of logic required for these deductions is precisely the ordinary quantificationa logic.

On the other hand, set theory is involved in logic itself, in several ways. For example, when studying a system of logic so as to establish its properties, we often find it useful to appeal to principles of set theory. Thus in the first chapter, when we were proving the completeness of an axiomatization of truth-functional logic, we found it convenient to define a certain sequence $A_1, A_2, A_3, \ldots$ of sets of formulae, and then to consider the union $\bigcup A_i (i \geqslant 1)$ of this sequence. Here we were profiting from several principles of set theory, including the axiom of union, so as to be able to affirm the existence of such a set. Actually, it is possible to prove the completeness of such an axiomatization of truth-functional logic by other methods that make a much more modest, and almost invisible, use of set theory, but the bolder method gives a proof of particular elegance.

There are other ways in which set theory is involved in logic. As we observed in the second chapter, in quantificational logic set theory appears in the definition of an interpretation, and thus indirectly in the definition of the relation of truistic implication. Here again it is possible to proceed more circumspectly. It is possible to define the concept of an interpretation and the relation of truistic implication in a way that makes less extensive use of the notions of set theory; but whatever definitions are followed, some such notions are still needed. Again, as we saw in the first chapter, the definition of the theses of an axiomatic system, if carried out rigorously, makes use of the idea of a set, or alternatively of various ideas of arithmetic. Indeed, the definition of what is to count as a formula of propositional or quantificational logic involves the same ideas, as was brought out in exercise 26. For that matter, even the concept of a solitary propositional letter involves the idea of a set. A propositional letter is not merely a mark on paper, but at best a *set* of marks of a single shape or form. Thus, for example, we say that the expression $(p \wedge q) \rightarrow p$ contains two distinct occurrences of the propositional letter p, and only one occurrence of the letter q. Although only two propositional letters occur in the expression, one of them occurs twice.

In view of these observations we can ask whether there is any sense in which logic is 'more fundamental' than set theory and other parts of mathematics such as arithmetic. The answer can perhaps be put as follows. Logic seems to be involved in set theory and arithmetic on the level of *use*; logic is an instrument within mathematics. On the other hand, set theory seems to be involved in logic on the level of *mention*. It is not involved in stating any particular principle of logic, or in deriving one principle from another. It is instead needed for making general statements *about* systems of logic – for setting out definitions of concepts such as 'propositional letter', 'formula', 'thesis', and 'interpretation', and for proving theorems about the interrelations of these concepts. In this respect it seems fair to say that logic is 'more fundamental' than either set theory or arithmetic.

3 *The nature of set theory*

Questions of priority also arise within set theory itself. Roughly speaking, two distinct traditions can be discerned in literature on the subject. On one tradition, the axioms of any theory of sets, say that of Zermelo and Fraenkel, are taken as a whole as being of more or less equal status, and the theorems of the theory are developed in the most elegant way attainable from the entire basis. On another tradition, the axioms of set theory are arranged in some kind of hierarchy, some being regarded as more fundamental, or more certain, or less dubious than others. Usually the axiom of infinity, the axiom scheme of replacement, the axiom of choice, and axioms of strong infinity are placed quite late in the list. Theorems are proven with this loose ordering in mind. The mathematician working in this tradition seeks to obtain the maximum possible with the least possible means, and a long proof using only the more secure principles is preferred to a shorter or more direct one that appeals to principles higher in the hierarchy.

The latter tradition has been elaborated in some detail by W. V. Quine who, moreover, has also considered its bearing upon the exposition of classical logic itself. One of the central questions to which Quine attends in his writings on logic is that of determining how much of set theory or arithmetic is required, as an absolute minimum, for the definition and basic study of ordinary quantificational logic. Pursuit of this problem has led him to define many of the concepts of logic, such as the relation of truistic implication, in ways more economical of sets and of principles about sets than those used in the earlier chapters of this book.

Such an approach to set theory would make good sense if the principles of that theory committed us to the existence of special entities of some kind, called sets, and distinct from such physical entities as heaps and piles as well as from such psychological entities as thoughts and concepts. For if a proposition does assert the existence of something, such as the empty set or an infinite set, then it is perfectly natural to ask for the evidence that there is for making the assertion, and it is also natural to order the axioms of set theory, which are largely existential or conditionally existential in their form, according to the confidence that we can muster in their truth. For example, the axiom of pairs might be accepted with considerable confidence. As we have seen, that axiom says merely that for any things x and y there is a set whose elements are just x and y. The axiom of powersets and the axiom of union might be given a little less confidence, and the axiom of infinity, say, might be regarded as very dubious indeed. Seen from this point of view, the minimization of the principles involved in a proof becomes a minimization of ontological commitment.

However, this is not the only way of understanding set theory. There is another possible view that emerges when we consider the case of set theory alongside that of other mathematical theories, such as geometry. In geometry there are existence statements and conditional existence statements in abundance, and yet, as was suggested in the fourth chapter, we scarcely feel tempted to see these statements as describing a special world of dimensionless points and perfectly straight lines. We see them instead as setting out a kind of structure, to which very good approximations may be found in the everyday world. A similar conception can be suggested for set theory. Its principles, and in particular its basic axioms, can be seen as setting out a pattern of relationships. The pattern is useful in so far as we can find interesting approximations to it in the everyday world, and exact exemplifications of it in *other* patterns of mathematics that already have an interest of their own. We can find approximations to the patterns of set theory in our processes of labelling and conceptualizing. When we sort mail, classify books, distinguish situations, then the wooden boxes, filing cards, and concepts that we use are interrelated in ways that approximate the structures of set theory. We can also find exact exemplifications of the patterns of set theory throughout classical mathematics. Concepts such as those of a function, an operation and the various kinds of number, as well as more recent notions such as those of a group or a topological space, can be seen as exemplifications of basic patterns provided by set theory. In other words, a function can be treated as a special kind of set, as can an operation, and similarly with the other concepts mentioned.

Such a view has several interesting applications. If the principles of set theory are patterns rather than assertions, and are neither true nor false, then in one respect even the most elementary of its axioms, such as the axiom of pairs, are in the same position as the most sophisticated among them, such as the axioms of strong infinity. Neither is true, neither is false, and so neither one is more dubious than the other. However there are other respects in which they differ. Some of the basic axioms of set theory, such as those of the empty set, pairs, powersets and union can be related quite directly to the conceptual processes that people use in the everyday world. But when we enter the higher domains of set theory, other considerations also become important. Some axioms, such as those of powersets and replacement, may find their principal justification in providing a pattern to which *other* patterns, from other parts of mathematics, may fit. Other axioms, such as that of choice, may be useful mainly as simplifying devices that make the structure of set theory more elegant, more manageable, and more easily explored.

This in turn leads to a principle of tolerance which emerges, for example, in connection with what is known as the *generalized continuum hypothesis*. A set x is said to be of lesser cardinality than a set y iff there is a one-one correspondence between x and some subset of y, but no one-one correspondence between y and any subset of x. According to the generalized continuum hypothesis, if x is any infinite set and y is its powerset, then there is no set z such that x is of lesser cardinality than z but z is of lesser cardinality than y.

Exercise 113 Explain, with the help of a simple example, why the generalized continuum hypothesis is formulated only for infinite sets.

Now it is known that the generalized continuum hypothesis is independent of the usual axioms of Zermelo-Fraenkel set theory, in the sense that neither it nor its negation is truistically implied by those axioms. It is also known that the addition of the generalized continuum hypothesis to the usual axioms serves as a simplifying device for some purposes. On the other hand, it may turn out that the addition of some *other* principle, incompatible with the continuum hypothesis, is also useful for other purposes that are internal to set theory. However, neither the continuum hypothesis nor its negation seems to have any advantage as a focus for approximations from the everyday world. In such a situation we would have two set theories incompatible with each other, but each of interest for certain purposes. There would be no necessity, indeed no possibility, of choosing one of the two as true and discounting the other as false.

The principle of tolerance applies further. We may consider not only various different ways of extending Zermelo-Fraenkel set theory, but also various alternative systems incompatible with even the most basic axioms of Zermelo. Such a set theory is the 'stratified' set theory of Quine. This system contains an axiom scheme of 'stratified comprehension' that enables it to assert the existence of, say, the set of all sets. If this is coupled with the axiom scheme of separation from the system of Zermelo, it leads quickly to a contradiction, but when it is considered in its own right it appears to be free from inconsistency. Now Quine's system provides a pattern for which we can find as good approximations in everyday life, and as good exemplifications in other parts of mathematics, as we can find for the Zermelo-Fraenkel system. The main reason for preferring the latter to the former is that the stratified set theory contains certain complexities which, in detailed applications, can be rather irritating. In particular, it can be quite annoying to have to check, upon each application of the scheme of stratified comprehension, that the comprehending formula of the scheme, when written out fully in primitive notation, really is stratified in the sense required.

These considerations raise a further possibility. We have suggested that set theory finds one of its uses as a pattern for approximating in the processes of sorting, labelling, and conceptualizing, and these are dynamic processes that involve change and take place over time. We may speculate whether it might be of interest to construct a dynamic set theory, or rather a theory of *dynamic sets:* a theory that recognizes change in that it allows a set to grow and even to diminish in membership, and perhaps also allows for sets to come into and pass out of existence.

Answers to selected exercises

Chapter 1

Exercise 1   We may take ⌐ and ∧ as primitive, and consider $\alpha \vee \beta$ as an abbreviation for ⌐(⌐$\alpha \wedge$ ⌐β). We may also take ⌐ and ∨ as primitive, and regard $\alpha \wedge \beta$ as an abbreviation for ⌐(⌐$\alpha \vee$ ⌐ β).

Exercise 2 (first part)   Suppose that α is not a tautology. Then by the definition of a tautology, there is an assignment of truth values to propositional letters that makes α come out as false. Clearly, this assignment makes ⌐α come out as true, and so ⌐α is not a counter-tautology. The converse is similar.

Exercise 5 (first part)   We may obtain $\alpha \wedge \beta \to \beta \wedge \alpha$ by the following derivation:

   (1) $\alpha \wedge \beta \to \beta$ axiom
   (2) $\alpha \wedge \beta \to \alpha$ axiom
   (3) $\alpha \wedge \beta \to \beta \wedge \alpha$ from (1) and (2) by rule of conjunction.

Exercise 11   Imagine the task completed, with just two litres in, say, the nine-litre bucket. We could obtain that if we could dispose of just seven litres from a full nine-litre bucket. We could do that if we could dispose of just three litres from a full nine-litre bucket. We could do that if we could have just one litre in the four-litre bucket . . . . And we can easily do *that*, by pouring off four litres from a full nine-litre bucket, then another four, and transferring the remaining one litre.

Exercise 15   An answer to the first part of this exercise is set out in section 1 of chapter 2. For the second part, the essential idea is to take $\beta \land \neg\beta \to \neg\alpha$ as already established, and apply the rule of contraposition, the principle of de Morgan, and various other axioms and rules.

Exercise 17 By the principle of induction we need only show two things: *first*, that the number 1 has the property, in other words that $1 = 1^2$ (this is the basis), and *second*, that whenever a positive integer n has the property, then $n + 1$ has it too (this is the induction step). Now the basis is immediately true. As for the induction step, we suppose that n has the property, in other words that $1 + 3 + \cdots + (2n - 1) = n^2$ (this is the induction hypothesis). Then

$$1 + 3 + \cdots + 2(n + 1) - 1 = (1 + 3 + \cdots + (2n - 1)) + 2(n + 1) - 1$$

and so, applying the induction hypothesis to the material in parentheses,

$$= n^2 + 2(n + 1) - 1$$
$$= n^2 + 2n + 1$$
$$= (n + 1)^2.$$

Exercise 20 (first part) The set of all formulae that are not contingent; the empty set; the set whose only element is $p \lor \neg p$; the set of all formulae; again the set of all formulae; the empty set; the set of all tautologies; the set whose only element is $p \lor \neg p$; the set whose elements are just $p \lor \neg p$ and all the countertautologies; the empty set.

Exercise 21 (first part) To show that $X \cap Y \subseteq X$ we need only show that every element of $X \cap Y$ is an element of X. But we know that an object is an element of $X \cap Y$ iff it is an element both of X and of Y.

Exercise 22 The set of all formulae that contain exactly one propositional letter; the set of all formulae that contain exactly two distinct propositional letters; the set of all formulae that contain at most two distinct propositional letters; the empty set; the set of all formulae; the set of all formulae that contain at least five distinct propositional letters.

Exercise 27 (in part)   We give the verification for the rule of contraposition. We need to show that whenever α tautologically implies β, then $\neg\beta$ tautologically implies $\neg\alpha$. Let α and β be formulae, and suppose that $\neg\beta$ does not tautologically imply $\neg\alpha$. Then there is an assignment of truth values under which $\neg\beta$ is true and $\neg\alpha$ is false. Clearly this assignment makes α true and β false. Thus α does not tautologically imply β.

Exercise 31   Suppose that X axiomatically yields ψ. Then there are formulae $\varphi_1, \ldots, \varphi_n$ in X such that $\varphi_1 \wedge \cdots \wedge \varphi_n \to \psi$ is a thesis. But since X is a subset of Y, each of $\varphi_1, \ldots, \varphi_n$ is also an element of Y, and so Y axiomatically yields ψ.

Exercise 34   Suppose that $\varphi \wedge \psi$ is an element of A. Then since each of $\varphi \wedge \psi \to \varphi$ and $\varphi \wedge \psi \to \psi$ is a thesis, A axiomatically yields each of φ and ψ, so that by claim (5), φ and ψ are both elements of A. Suppose, conversely, that each of φ and ψ is an element of A. Then since $\varphi \wedge \psi \to \varphi \wedge \psi$ is a thesis, A axiomatically yields $\varphi \wedge \psi$, and so by claim (5), $\varphi \wedge \psi$ is an element of A.

Chapter 2

Exercise 37 (first part)   Let α and β be formulae and suppose that β is a tautology. Then there is no assignment of truth values to propositional letters that makes β come out false. Hence, *a fortiori*, there is no assignment under which β comes out false whilst α comes out true.

Exercise 40   The expression $p \wedge \neg p \to (p \vee q) \wedge \neg p$ is a substitution instance of the expression $p \wedge r \to (p \vee q) \wedge r$, which has the appropriate properties. Also, the expression $(p \vee q) \wedge \neg p \to q$ is a substitution instance of itself, which has the appropriate properties.

Exercise 42   If p and q are distinct propositional letters, then $(p \wedge q) \vee (\neg p \vee \neg q)$ is subtended by each of p and q considered separately, but not by their disjunction $p \vee q$.

Exercise 43 $v(\alpha \vee \beta) = 1$.

Exercise 44 $v(\alpha \wedge \beta) = a$, $v(\alpha \vee \beta) = 1$.

Exercise 48   Suppose that α and β have no propositional letters in common. Then we can assign the value a to all the propositional letters in α, whilst assigning the value b to all the propositional letters in β. It is easy to verify by an inductive argument that for such an assignment v, we have $v(\alpha) = a$ whilst $v(\beta) = b$, so that $v(\alpha) \not\leqslant v(\beta)$, and thus the expression $\alpha \rightarrow \beta$ is not valid in the four element model.

Exercise 55   To abandon the law of identity would amount, in the present context, to allowing a single propositional letter to be assigned distinct values at its various occurrences within a single formula. It can be shown that this leads to a logic with no theses at all.

Chapter 3

Exercise 57   A formula of the form $\forall x(\alpha)$ is false under an interpretation iff α is false under at least one x-variant of that interpretation.

Exercise 58   A formula of the form $\exists x(\alpha)$ is true under an interpretation iff α is true under at least one x-variant of that interpretation.

Exercise 59   Choose a domain of discourse with just two elements, a and b. Put $F(P) = \{a\}$, $F(Q) = \{b\}$, and $f(x) = a$. Call this interpretation J. Note that J has just two x-variants: itself and the interpretation J' obtained by putting $f'(x) = b$. Note that $Px \vee Qx$ is true under each of these two x-variants, so that $\forall x(Px \vee Qx)$ is true under J. On the other hand, Px is false under J', so that $\forall x(Px)$ if false under J. Similarly Qx is false under J, so that $\forall x(Qx)$ is also false under J. Thus $\forall x(Px) \vee \forall x(Qx)$ is false under J.

Exercise 62   Suppose that $\forall x(\alpha)$ does not truistically imply α. Then there is an interpretation J under which $\forall x(\alpha)$ comes out true whilst α comes out false. Since $\forall x(\alpha)$ is true under J, we have that α is true under every x-variant of J, and in particular, under J itself. This gives us a contradiction.

Exercise 67 All the occurrences of x are bound. However the first two occurrences of z are free, as also are the last two occurrences of y.

Exercise 69 The first substitution is irregular, the second is regular.

Exercise 71 The first two substitutions are vacuous, and so are regular. The remainder are all irregular.

Exercise 72 From $\alpha \rightarrow \beta$ to $\alpha \rightarrow \forall x(\beta)$, where α and β are formulae and where x does not have any free occurrences in α.

Exercise 73 (first part)

(1)	$\forall x(\alpha \wedge \beta) \rightarrow \alpha \wedge \beta$	axiom
(2)	$\alpha \wedge \beta \rightarrow \alpha$	axiom
(3)	$\forall x(\alpha \wedge \beta) \rightarrow \alpha$	(1) and (2) by transitivity
(4)	$\forall x(\alpha \wedge \beta) \rightarrow \forall x(\alpha)$	(3) by rule for $\forall$
(5)	$\alpha \wedge \beta \rightarrow \beta$	axiom
(6)	$\forall x(\alpha \wedge \beta) \rightarrow \beta$	(1) and (5) by transitivity
(7)	$\forall x(\alpha \wedge \beta) \rightarrow \forall x(\beta)$	(6) by rule for $\forall$
(8)	$\forall x(\alpha \wedge \beta) \rightarrow \forall x(\alpha) \wedge \forall x(\beta)$	(4) and (7) by rule for conjunction

Exercise 84 If n is even, put $f(n) = n/2$, and if n is odd, put $f(n) = -(n-1)/2$.

Exercise 86 We can define $x = y$ to be an abbreviation for $\forall z(Rxz \equiv Ryz) \wedge \forall z(Rzx \equiv Rzy)$.

Exercise 88 We would need to define identity as indistinguishability with respect to all the predicate letters, and this would require an infinitely long formula. Ordinary quantificational logic does not have any infinitely long formulae.

Chapter 4

Exercise 91 (in part) We construct a derivation for the quantifier negation principle $\exists x(\neg\alpha) \rightarrow \neg\forall x(\alpha)$:

(1) $\neg\alpha \land \forall x(\alpha) \rightarrow \forall x(\alpha)$ axiom
(2) $\forall x(\alpha) \rightarrow \alpha$ axiom
(3) $\neg\alpha \land \forall x(\alpha) \rightarrow \alpha$ (1) and (2) by transitivity
(4) $\neg\alpha \land \forall x(\alpha) \rightarrow \neg\alpha$ axiom
(5) $\neg\alpha \land \forall x(\alpha) \rightarrow \alpha \land \neg\alpha$ (3) and (4) by conjunction
(6) $\neg\alpha \rightarrow \neg\forall x(\alpha)$ (5) by *reductio ad absurdum*
(7) $\exists x(\neg\alpha) \rightarrow \neg\forall x(\alpha)$ (6) by rule for $\exists$

Exercise 92 Suppose that $\alpha \rightarrow \beta$ is a thesis. Then there is a derivation that leads to $\alpha \rightarrow \beta$. This derivation may be continued as follows:

(1) $\alpha \rightarrow \beta$ thesis
(2) $\neg\beta \land \alpha \rightarrow \alpha$ axiom
(3) $\neg\beta \land \alpha \rightarrow \beta$ (1) and (2) by transitivity
(4) $\neg\beta \land \alpha \rightarrow \neg\beta$ axiom
(5) $\neg\beta \land \alpha \rightarrow \beta \land \neg\beta$ (3) and (4) by conjunction
(6) $\neg\beta \rightarrow \neg\alpha$ (5) by *reductio ad absurdum*

Exercise 93 We can use exactly the same derivation as we did for this principle in classical propositional logic.

Exercise 94 The formula $(p \supset q) \lor (q \supset p)$ is intuitionistically valid only if for any statements α and β there is a construction π that is a proof of $(\alpha \supset \beta) \lor (\beta \supset \alpha)$. For this to be so, π would have to be a proof of $\alpha \supset \beta$ or a proof of $\beta \supset \alpha$, and for this to be so, π would have to contain a construction σ, and also contain either a proof that σ transforms any proof of α into a proof of β, or else a proof that σ transforms any proof of β into a proof of α. However, there does not seem to be any convincing reason why such a π and σ should exist.

Chapter 5

Exercise 99 If x and y are both empty sets, then they have exactly the same elements, and so by the axiom of extensionality they are identical.

Exercise 100   Take y to be x in the axiom of pairs.

Exercise 101   Let x and y be sets. By the axiom of pairs there is a set w whose elements are just x and y. Hence by the axiom of union there is a set z whose elements are just the things that are elements of at least one of the elements of w. Clearly z is the desired set.

Exercise 102   Let x be any non-empty set. By the axiom of union, there is a set w whose elements are just those things that are elements of at least one of the elements of x. Hence by the axiom scheme of separation, there is a set y whose elements are just the things z in w that satisfy the condition 'z is an element of *every* element of x'. Clearly y is the desired set.

Exercise 104   Suppose for *reductio ad absurdum* that there is a set y such that every set is an element of y. Then we may apply the axiom scheme of separation to obtain a set w whose elements are just the things z that are elements of y and satisfy the condition 'z is not an element of itself'. In other words, w is the set of all sets that are not elements of themselves. But we already know how this leads to a contradiction.

Exercise 108   If x has an element y that is empty, then clearly no element of y can be an element of z.

Exercise 109   Consider the case where $x = \{y_1, y_2, y_3\}$, and where $y_1 = \{a, b\}, y_2 = \{b, c\}, y_3 = \{c, a\}$, where a, b, c are three distinct objects.

Exercise 113   It is not difficult to show that if x has n elements, where n is a positive integer, then the powerset of x has 2^n elements. In particular, if x has 2 elements, then its powerset has 4 elements, and so any set with just 3 elements will have an intermediate cardinality.

Guide to further reading

This guide is drawn up to assist the student of philosophy, with a limited background in mathematics, in reading further in modern logic. It is very brief and selective.

General

Among the more general books, we mention A. N. Prior *Formal Logic* 2nd ed. (Oxford, Clarendon Press, 1962), W. V. Quine *Philosophy of Logic* (Englewood Cliffs, N.J., Prentice Hall, 1970), P. F. Strawson *Introduction to Logical Theory* (London, Methuen, 1952), and W. and M. Kneale *The Development of Logic* (Oxford, Clarendon Press, 1962). Each of these books is written for the undergraduate student in philosophy, but the interests of the authors are quite different. Prior's main interests are in nonclassical propositional logics, Quine gives particular attention to the relations between logic, set theory and ontology, Strawson is much concerned with the links and contrasts between formal logic and everyday uses of language, whilst the Kneales describe the history of the subject.

G. T. Kneebone *Mathematical Logic and the Foundations of Mathematics* (New York, Van Nostrand, 1963) is rather more mathematical in its approach than the others so far mentioned. Even more so is E. Mendelson *Introduction to Mathematical Logic* (New York, Van Nostrand, 1964).

Those wishing to search the journals for papers on particular subjects are advised to use the indexes and abstracts of *The Philosopher's Index.* This covers material from 1967 onwards. Papers on mathematical aspects of logic are reviewed regularly in the *Journal of Symbolic Logic, Mathematical Reviews,* and *Zentralblatt für Mathematik.*

Chapter 1

Those interested in questions of heuristics, and in particular the distinction between working forwards and working backwards, should read G. Polya *How To Solve It* (Princeton, N.J., Princeton University Press, 1945), especially the sections entitled 'Pappus' and 'Working Backwards'.

For further background on tools from set theory see A. A. Fraenkel *Abstract Set Theory* 2nd ed. (Amsterdam, North Holland, 1961), or for a very elementary treatment, S. Lipschutz *Set Theory and Related Topics* (New York, Schaum, 1964).

Chapter 2

The student interested in the problem of entailment could begin by reading T. F. Smiley 'Entailment and Deducibility' *Aristotelian Society Proceedings*, New Series, vol. 59 (1958-9), 233-54, remembering however that this paper was written in ignorance of the relation of de Morgan implication. He could also read A. R. Anderson and N. D. Belnap Jr. 'Tautological Entailments' *Philosophical Studies*, vol. 13 (1961), 9-24. However some care is needed with terminology. The relation that Anderson and Belnap call 'tautological entailment' is *not* the same as the one that we have been calling 'tautological implication'. It coincides instead with de Morgan implication, and its very purpose is to provide a contrast with the classical, or tautological, implication.

The relation of subtending is defined, without being given a special name, by T. F. Smiley on page 240 of the paper mentioned above. The idea is also taken up by P. T. Geach 'Entailment' *The Philosophical Review*, vol. 79 (1970), 237-9. The concept of tautopical implication is more or less implicit in a brief report by W. T. Parry in *Ergebnisse Eines Mathematischen Kolloquiums*, Heft 4, for 1931-2 (published 1933), 5-6, as also in the comments of Gödel on page 6 of the same *Ergebnisse*, and also in W. T. Parry 'The logic of C. I. Lewis' in P. A. Schilpp (ed.) *The Philosophy of C. I. Lewis* (Illinois, Open Court, 1968), especially pages 150-3.

However Parry's direct concern there is not the relation of tautopical implication itself, but rather a closely associated operator.

For an introduction to the attempt to transform de Morgan implication into an interesting operator, see A. R. Anderson 'Some unsolved problems concerning the system E of entailment' *Acta Philosophical Fennica*, vol. 16 (1963), 7–13. Some of Anderson's open problems have since been solved, particularly in a very technical paper by R. K. Meyer and J. M. Dunn in the *Journal of Symbolic Logic*, vol. 34 (1969), 460–74. A. R. Anderson and N. D. Belnap Jr. are planning to publish a comprehensive book on entailment in the near future.

A good place to begin reading about modal logic is G. E. Hughes and M. J. Cresswell *An Introduction to Modal Logic* (London, Methuen, 1968). Written for the student with a limited mathematical background, this book surveys the field and contains a comprehensive bibliography. There are also some interesting comments on present trends in research on modal logic and related subjects in D. Scott 'Advice on modal logic', in K. Lambert (ed.) *Philosophical Problems in Logic* (Dordrecht, Reidel, 1970).

Chapter 3

There is more material on the formal properties of quantificational logic in E. Mendelson *Introduction to Mathematical Logic* (New York, Van Nostrand, 1964). However, to avoid confusions the reader should note that the definition that we have been using of the truth of a formula under an interpretation differs marginally from that used by Mendelson. In effect, Mendelson uses two concepts, of 'truth' and of 'satisfaction', and our 'truth' corresponds to his 'satisfaction'. A variant account of the quantifiers is discussed in J. M. Dunn and N. D. Belnap Jr. 'The substitution interpretation of the quantifiers' *Nous*, vol. 2 (1968), 177–85. Questions concerning the structure of derivations in quantificational logic are studied in a classic paper of G. Gentzen, translated as 'Investigations into logical deduction' *American Philosophical Quarterly*, vol. 1 (1964), 288–306, and vol. 2 (1965), 204–18, and more recently in R. M. Smullyan *First-Order Logic* (Berlin, Springer-Verlag, 1968).

In recent years, the compactness theorem has been used to provide a coherent basis for the notion of 'infinitely small numbers', introduced in the early days of the differential and integral calculus and then discredited in the nineteenth century. For a general view of the subject, compare B. Russell *The Principles of Mathematics* (London, Allen and Unwin, 1903), chapters 39 and 40, with A. Robinson 'The metaphysics of the calculus', in I. Lakatos *Problems in the Philosophy of Mathematics* (Amsterdam, North Holland, 1967), reprinted in J. Hintikka (ed.) *The Philosophy of Mathematics* (London, Oxford University Press, 1969).

The Skolem paradox is discussed by T. Skolem himself in his paper 'Some remarks on axiomatic set theory' translated in J. van Heijenoort (ed.) *From Frege to Gödel* (Cambridge, Mass., Harvard University Press, 1967), 113–17. Further articles by Skolem are listed in the bibliography of van Heijenoort's book. The Skolem paradox is also discussed by J. R. Myhill 'The ontological significance of the Löwenheim-Skolem theorem', in M. White (ed.) *Academic Freedom, Logic and Religion* (Philadelphia, Pa., University of Pennsylvania Press, 1951), reprinted in part in I. M. Copi and J. A. Gould (eds.) *Contemporary Readings in Logical Theory* (New York, Macmillan, 1967).

For a general introduction to questions of undecidability and incompleteness see G. T. Kneebone *Mathematical Logic and the Foundations of Mathematics* (New York, Van Nostrand, 1963). Some of these ideas are set out in a popular way by E. Nagel and J. R. Newman *Gödel's Proof* (New York, New York University Press, 1958). However, it should be borne in mind that the latter gives only a very preliminary orientation towards the incompleteness theorems. For example, it skirts around the concept of an effective procedure, which nevertheless plays an important part in the proof. There is a detailed technical account of the incompleteness theorems in Mendelson's book, mentioned above, as also in S. C. Kleene *Introduction to Metamathematics* (Amsterdam, North Holland, 1952).

For a brief survey of some ideas in the theory of effective procedures see H. Rogers Jr. 'The present theory of Turing machine computability' *Journal of the Society for Industrial and Applied Mathematics*, vol. 7 (1959), 114–30, reprinted in J. Hintikka (ed.) *The Philosophy of Mathematics* (London, Oxford University Press,

1969). There is an elementary introduction to the theory of
recursive functions in A. Grzegorczyk *Fonctions récursives* (Paris,
Gauthier-Villars, 1961).

There is a general discussion of the expressive capacities of
quantificational logic in W. V. Quine *Philosophy of Logic* (Englewood Cliffs, N.J., Prentice Hall, 1970). Some questions concerning
the use of adjectival constructions in ordinary language arise in
J. Austin *Sense and Sensibilia* (London, Oxford University Press,
1962), section 7; J. Bennett 'Real' *Mind*, vol. 75 (1966), 501-15;
and R. Clark 'Concerning the logic of predicate modifiers' *Nous*,
vol. 4 (1970), 311-35.

Chapter 4

The basic ideas of intuitionism were first set out by the Dutch
mathematician L. E. J. Brouwer, and several of his papers are
available in English: see J. van Heijenoort (ed.) *From Frege to
Gödel* (Cambridge, Mass., Harvard University Press, 1967), and
P. Benacerraf and H. Putnam (eds.) *Philosophy of Mathematics:
Selected Readings* (Englewood Cliffs, N.J., Prentice Hall, 1964).
These papers are of particular interest for the general philosophical
ideas underlying intuitionism. There is also a careful description of
Brouwer's ideas by G. Kreisel and M. Newman 'Luitzen Egbertus
Jan Brouwer (1881-1966)' *Biographical Memoirs of Fellows of the
Royal Society*, vol. 15 (1969), 39-68, published by The Royal
Society, London. For systematic expositions of the formal aspects
of intuitionistic mathematics and logic, see A. Heyting *Intuitionism,
an Introduction* (Amsterdam, North Holland, 1956), and
A. S. Troelstra *Principles of Intuitionism*, Lecture Notes in
Mathematics, vol. 95 (Berlin, Springer-Verlag, 1969). Some further
ideas of revelance to intuitionism are explored by M. A. E. Dummett
'Truth' *Aristotelian Society Proceedings*, New Series, vol. 59
(1958-9), 141-6, reprinted in P. F. Strawson (ed.) *Philosophical
Logic* (London, Oxford University Press, 1967).

Chapter 5

For a general background in set theory, see A. A. Fraenkel
Abstract Set Theory 2nd ed. (Amsterdam, North Holland, 1961).

There are many different approaches to the axiomatization of set theory, and some of the main ones are surveyed in H. Wang and R. McNaughton *Les systèmes axiomatiques de la théorie des ensembles* (Paris, Gauthier-Villars, 1953), with further material in W. S. Hatcher *Foundations of Mathematics* (Philadelphia, Saunders, 1968), and in W. V. Quine *Set Theory and its Logic* (Cambridge, Mass., Harvard University Press, 1963), part 3.

In the first two parts of the same book, Quine works on his programme of minimizing the ontological commitments of set theory, and in his more recent *Philosophy of Logic* (Englewood Cliffs, N.J., Prentice Hall, 1970), he discusses the bearing of such a programme on the formulation of ordinary quantificational logic itself. In this connection, see also H. Putnam *Philosophy of Logic* (New York, Harper, 1971).

There is some discussion of the nature of set theory in A. A. Fraenkel and Y. Bar-Hillel *Foundations of Set Theory* (Amsterdam, North Holland, 1958; new ed., 1972). Kurt Gödel has presented his ideas on the subject in an article 'What is Cantor's Continuum Problem?' *American Mathematical Monthly*, vol. 54 (1947), 515–25, reprinted with a postscript in P. Benacerraf and H. Putnam (eds.) *Philosophy of Mathematics: Selected Readings* (Englewood Cliffs, N.J., Prentice Hall, 1964). Gödel's ideas are further explained and discussed in a very clear article by S. Barker 'Realism as a philosophy of mathematics' in J. Bulloff and others (eds.) *Foundations of Mathematics* (Berlin, Springer-Verlag, 1969). Quite different ideas on set theory are set out by M. Black 'The elusiveness of sets' *Review of Metaphysics*, vol. 24 (1971), 614–36. Views more akin to those of the present book are expressed by P. J. Cohen and R. Hersch 'Non-Cantorian set theory' *Scientific American*, vol. 217 (December 1967), 104–16, and by A. Rényi *Dialogues on Mathematics* (San Francisco, Holden-Day, 1967).

Index

This does not include entries in the guide to further reading.